Edited by
Tharwat F. Tadros

Emulsion Formation and Stability

Related Titles

Tadros, T. F.

Dispersion of Powders

in Liquids and Stabilization of Suspensions

2012

ISBN: 978-3-527-32941-0

Tadros, T. F. (ed.)

Self-Organized Surfactant Structures

2010

ISBN: 978-3-527-31990-9

Tadros, T. F. (ed.)

Colloids and Interface Science Series

6 Volume Set

2010

ISBN: 978-3-527-31461-4

Tadros, T. F.

Rheology of Dispersions

2010

ISBN: 978-3-527-32003-5

Wilkinson, K. J., Lead, J. R. (eds.)

Environmental Colloids and Particles

Behaviour, Separation and Characterisation

2007

ISBN: 978-0-470-02432-4

Edited by Tharwat F. Tadros

Emulsion Formation and Stability

WILEY-VCH Verlag GmbH & Co. KGaA

The Editor

Prof. Dr. Tharwat F. Tadros
89 Nash Grove Lane
Wokingham
Berkshire RG40 4HE
United Kingdom

All books published by **Wiley-VCH** are carefully produced. Nevertheless, authors, editors, and publisher do not warrant the information contained in these books, including this book, to be free of errors. Readers are advised to keep in mind that statements, data, illustrations, procedural details or other items may inadvertently be inaccurate.

Library of Congress Card No.: applied for

British Library Cataloguing-in-Publication Data
A catalogue record for this book is available from the British Library.

Bibliographic information published by the Deutsche Nationalbibliothek
The Deutsche Nationalbibliothek lists this publication in the Deutsche Nationalbibliografie; detailed bibliographic data are available on the Internet at <http://dnb.d-nb.de>.

© 2013 Wiley-VCH Verlag GmbH & Co. KGaA, Boschstr. 12, 69469 Weinheim, Germany

All rights reserved (including those of translation into other languages). No part of this book may be reproduced in any form – by photoprinting, microfilm, or any other means – nor transmitted or translated into a machine language without written permission from the publishers. Registered names, trademarks, etc. used in this book, even when not specifically marked as such, are not to be considered unprotected by law.

Print ISBN: 978-3-527-31991-6
ePDF ISBN: 978-3-527-64797-2
ePub ISBN: 978-3-527-64796-5
mobi ISBN: 978-3-527-64795-8
oBook ISBN: 978-3-527-64794-1

Cover Design   Adam-Design, Weinheim
Typesetting   Laserwords Private Limited, Chennai, India
Printing and Binding   Markono Print Media Pte Ltd, Singapore

Contents

Preface *XI*
List of Contributors *XIII*

1 Emulsion Formation, Stability, and Rheology *1*
Tharwat F. Tadros
1.1 Introduction *1*
1.1.1 Nature of the Emulsifier *1*
1.1.2 Structure of the System *2*
1.1.3 Breakdown Processes in Emulsions *3*
1.1.4 Creaming and Sedimentation *3*
1.1.5 Flocculation *4*
1.1.6 Ostwald Ripening (Disproportionation) *4*
1.1.7 Coalescence *4*
1.1.8 Phase Inversion *4*
1.2 Industrial Applications of Emulsions *4*
1.3 Physical Chemistry of Emulsion Systems *5*
1.3.1 The Interface (Gibbs Dividing Line) *5*
1.4 Thermodynamics of Emulsion Formation and Breakdown *6*
1.5 Interaction Energies (Forces) between Emulsion Droplets and Their Combinations *8*
1.5.1 van der Waals Attraction *8*
1.5.2 Electrostatic Repulsion *9*
1.5.3 Steric Repulsion *12*
1.6 Adsorption of Surfactants at the Liquid/Liquid Interface *14*
1.6.1 The Gibbs Adsorption Isotherm *14*
1.6.2 Mechanism of Emulsification *17*
1.6.3 Methods of Emulsification *19*
1.6.4 Role of Surfactants in Emulsion Formation *21*
1.6.5 Role of Surfactants in Droplet Deformation *22*
1.7 Selection of Emulsifiers *26*
1.7.1 The Hydrophilic–Lipophilic Balance (HLB) Concept *26*
1.7.2 The Phase Inversion Temperature (PIT) Concept *29*
1.7.3 The Cohesive Energy Ratio (CER) Concept *31*

1.7.4 The Critical Packing Parameter (CPP) for Emulsion Selection *32*
1.8 Creaming or Sedimentation of Emulsions *35*
1.8.1 Creaming or Sedimentation Rates *36*
1.8.2 Prevention of Creaming or Sedimentation *37*
1.9 Flocculation of Emulsions *40*
1.9.1 Mechanism of Emulsion Flocculation *40*
1.9.1.1 Flocculation of Electrostatically Stabilized Emulsions *41*
1.9.1.2 Flocculation of Sterically Stabilized Emulsions *42*
1.9.2 General Rules for Reducing (Eliminating) Flocculation *43*
1.10 Ostwald Ripening *44*
1.11 Emulsion Coalescence *45*
1.11.1 Rate of Coalescence *46*
1.11.2 Phase Inversion *47*
1.12 Rheology of Emulsions *48*
1.12.1 Interfacial Rheology *48*
1.12.1.1 Interfacial Tension and Surface Pressure *48*
1.12.1.2 Interfacial Shear Viscosity *49*
1.12.2 Measurement of Interfacial Viscosity *49*
1.12.3 Interfacial Dilational Elasticity *50*
1.12.4 Interfacial Dilational Viscosity *51*
1.12.5 Non-Newtonian Effects *51*
1.12.6 Correlation of Emulsion Stability with Interfacial Rheology *51*
1.12.6.1 Mixed Surfactant Films *51*
1.12.6.2 Protein Films *51*
1.13 Bulk Rheology of Emulsions *53*
1.13.1 Analysis of the Rheological Behavior of Concentrated Emulsions *54*
1.14 Experimental $\eta_r - \phi$ Curves *57*
1.14.1 Experimental $\eta_r - \phi$ Curves *58*
1.14.2 Influence of Droplet Deformability *58*
1.15 Viscoelastic Properties of Concentrated Emulsions *59*
1.15.1 High Internal Phase Emulsions (HIPEs) *61*
1.15.2 Deformation and Breakup of Droplets in Emulsions during Flow *66*
References *73*

2 Emulsion Formation in Membrane and Microfluidic Devices *77*
Goran T. Vladisavljević, Isao Kobayashi, and Mitsutoshi Nakajima
2.1 Introduction *77*
2.2 Membrane Emulsification (ME) *78*
2.2.1 Direct Membrane Emulsification *78*
2.2.2 Premix Membrane Emulsification *79*
2.2.3 Operating Parameters in Membrane Emulsification *80*
2.2.4 Membrane Type *80*
2.2.4.1 Surfactant Type *80*
2.2.4.2 Transmembrane Pressure and Wall Shear Stress *81*
2.3 Microfluidic Junctions and Flow-Focusing Devices *82*

2.3.1	Microfluidic Junctions *82*	
2.3.2	Microfluidic Flow-Focusing Devices (MFFD) *83*	
2.4	Microfluidic Devices with Parallel Microchannel Arrays *85*	
2.4.1	Grooved-Type Microchannel Arrays *86*	
2.4.2	Straight-through Microchannel Arrays *88*	
2.5	Glass Capillary Microfluidic Devices *89*	
2.6	Application of Droplets Formed in Membrane and Microfluidic Devices *93*	
2.7	Conclusions *93*	
	Acknowledgments *94*	
	References *94*	

3 Adsorption Characteristics of Ionic Surfactants at Water/Hexane Interface Obtained by PAT and ODBA *99*

Nenad Mucic, Vincent Pradines, Aliyar Javadi, Altynay Sharipova, Jürgen Krägel, Martin E. Leser, Eugene V. Aksenenko, Valentin B. Fainerman, and Reinhard Miller

3.1	Introduction *99*	
3.2	Experimental Tools *99*	
3.3	Theory *101*	
3.4	Results *102*	
3.5	Summary *107*	
	Acknowledgments *107*	
	References *107*	

4 Measurement Techniques Applicable to the Investigation of Emulsion Formation during Processing *109*

Nima Niknafs, Robin D. Hancocks, and Ian T. Norton

4.1	Introduction *109*	
4.2	Online Droplet Size Measurement Techniques *112*	
4.2.1	Laser Systems *112*	
4.2.2	Sound Systems *115*	
4.2.3	Direct Imaging *115*	
4.2.4	Other Techniques *118*	
4.3	Techniques Investigating Droplet Coalescence *121*	
4.4	Concluding Remarks *123*	
	References *125*	

5 Emulsification in Rotor–Stator Mixers *127*

Andrzej W. Pacek, Steven Hall, Michael Cooke, and Adam J. Kowalski

5.1	Introduction *127*	
5.2	Classification and Applications of Rotor–Stator Mixers *128*	
5.2.1	Colloid Mills *129*	
5.2.2	In-Line Radial Discharge Mixers *130*	
5.2.3	Toothed Devices *131*	

5.2.4	Batch Radial Discharge Mixers 132
5.2.5	Design and Arrangement 133
5.2.6	Operation 136
5.3	Engineering Description of Emulsification/Dispersion Processes 138
5.3.1	Drop Size Distributions and Average Drop Sizes 138
5.3.2	Drop Size in Liquid–Liquid Two-Phase Systems – Theory 140
5.3.3	Maximum Stable Drop Size in Laminar Flow 141
5.3.4	Maximum Stable Drop Size in Turbulent Flow 142
5.3.5	Characterization of Flow in Rotor–Stator Mixers 143
5.3.5.1	Shear Stress 143
5.3.5.2	Average Energy Dissipation Rate 144
5.3.5.3	Power Draw 144
5.3.6	Average Drop Size in Liquid–Liquid Systems 145
5.3.7	Scaling-up of Rotor–Stator Mixers 147
5.4	Advanced Analysis of Emulsification/Dispersion Processes in Rotor–Stator Mixers 152
5.4.1	Velocity and Energy Dissipation Rate in Rotor–Stator Mixers 153
5.4.1.1	Batch Rotor–Stator Mixers 154
5.4.1.2	In-Line Rotor–Stator Mixers 157
5.4.2	Prediction of Drop Size Distributions during Emulsification 160
5.5	Conclusion 163
	Nomenclature 163
	References 165

6 Formulation, Characterization, and Property Control of Paraffin Emulsions 169

Jordi Esquena and Jon Vilasau

6.1	Introduction 169
6.1.1	Industrial Applications of Paraffin Emulsions 170
6.1.2	Properties of Paraffin 170
6.1.3	Preparation of Paraffin Emulsions 172
6.2	Surfactant Systems Used in Formulation of Paraffin Emulsions 174
6.2.1	Phase Behavior 175
6.3	Formation and Characterization of Paraffin Emulsions 178
6.4	Control of Particle Size 181
6.5	Stability of Paraffin Emulsions 185
6.5.1	Stability as a Function of Time under Shear (Orthokinetic Stability) 185
6.5.2	Stability as a Function of Freeze–Thaw Cycles 186
6.5.3	Stability as a Function of Electrolytes 189
6.6	Conclusions 195
	Acknowledgments 196
	References 196

7		**Polymeric O/W Nano-emulsions Obtained by the Phase Inversion Composition (PIC) Method for Biomedical Nanoparticle Preparation** *199*
		Gabriela Calderó and Conxita Solans
7.1		Introduction *199*
7.2		Phase Inversion Emulsification Methods *200*
7.3		Aspects on the Choice of the Components *201*
7.4		Ethylcellulose Nano-Emulsions for Nanoparticle Preparation *202*
7.5		Final Remarks *204*
		Acknowledgments *205*
		References *205*
8		**Rheology and Stability of Sterically Stabilized Emulsions** *209*
		Tharwat F. Tadros
8.1		Introduction *209*
8.2		General Classification of Polymeric Surfactants *210*
8.3		Interaction between Droplets Containing Adsorbed Polymeric Surfactant Layers: Steric Stabilization *212*
8.3.1		Mixing Interaction G_{mix} *213*
8.3.2		Elastic Interaction G_{el} *214*
8.4		Emulsions Stabilized by Polymeric Surfactants *216*
8.4.1		W/O Emulsions Stabilized with PHS-PEO-PHS Block Copolymer *219*
8.5		Principles of Rheological Techniques *220*
8.5.1		Steady State Measurements *220*
8.5.1.1		Bingham Plastic Systems *221*
8.5.1.2		Pseudoplastic (Shear Thinning) System *221*
8.5.1.3		Herschel–Bulkley General Model *222*
8.5.2		Constant Stress (Creep) Measurements *222*
8.5.3		Dynamic (Oscillatory) Measurements *223*
8.6		Rheology of Oil-in-Water (O/W) Emulsions Stabilized with Poly(Vinyl Alcohol) *226*
8.6.1		Effect of Oil Volume Fraction on the Rheology of the Emulsions *226*
8.6.2		Stability of PVA-Stabilized Emulsions *229*
8.6.3		Emulsions Stabilized with an A-B-A Block Copolymer *236*
8.6.4		Water-in-Oil Emulsions Stabilized with A-B-A Block Copolymer *240*
		References *245*

Index *247*

Preface

This book is based on selection of some papers from the Fifth World Congress on Emulsions that was held in Lyon, in October 2010. These series of World congresses emphasize the importance of emulsions in industry, including food, cosmetics, pharmaceuticals, agrochemicals, and paints. Following each meeting, a number of topics were selected, the details of which were subsequently published in the journals, Colloids and Surfaces and Advances in Colloid and Interface Science. The selected papers of the fourth Congress (2006) were published by Wiley-VCH (Germany).

This book contains selected topics from the Fifth World Congress, the title of which "Emulsion Formation and Stability" reflects the importance of emulsification techniques, the production of nanoparticles for biomedical applications as well as the importance of application of rheological techniques for studying the interaction between the emulsion droplets.

Chapter 1 describes the principles of emulsion formation, selection of emulsifiers, and control of emulsion stability. A section is devoted to the rheology of emulsions, including both interfacial rheology as well as the bulk rheology of emulsions. Chapter 2 deals with emulsion formation using membrane and microfluidics devices. In membrane emulsification (ME), the system is produced by injection of a pure disperse phase or a premix of a coarse emulsion into the continuous phase through a microporous membrane. Hydrophobic membranes are used to produce water-in-oil (W/O) emulsions, whereas hydrophilic membranes are used to produce oil-in-water (O/W) emulsions. In microfluidics, the combined two-phase flow is forced through a small orifice that allows one to obtain monodisperse droplets. Chapter 3 deals with adsorption of ionic surfactants at the hexane/water interface using the profile analysis technique (PAT) and the oscillating drop and bubble analyzer (ODBA). Theoretical models were used to analyze the adsorption results. Chapter 4 describes the various techniques that can be applied to investigate emulsion formation during processing. The effect of different emulsion techniques on the droplet size distribution was investigated using various methods such as light diffraction and ultrasound. Particular attention was given to online droplet size measurements. Chapter 5 deals with emulsification using rotor–stator mixers that are commonly used in industry, both in laboratory and large-scale production of emulsions. The various types of rotor–stator mixers are described. The selection of a rotor–stator mixer for a specific end product depends on the required droplet size distribution and the scale of the process. Chapter 6 describes the formulation, characterization, and property control of paraffin emulsions. The industrial application of paraffin emulsions is described highlighting the property of paraffin and method of preparation. The surfactants used in formation of paraffin emulsions are described in terms of their phase behavior. The control of particle size and its distribution of the resulting emulsion are described at a fundamental level. Chapter 7 describes polymeric O/W nanoemulsions produced by the phase inversion composition (PIC) method with application of the resulting nanoparticles in biomedicine. A description of the PIC method is given with reference to the aspects of choice of the components. The production of ethyl cellulose nanoparticles is described. Chapter 8 gives a detailed analysis of the rheology and stability of sterically stabilized emulsions. A section is devoted to the general classification of polymeric surfactants followed by discussion of the theory of sterically stabilized emulsions. The application of block and graft copolymers for preparation of highly stable emulsions is described. The principles of the various rheological techniques that can be applied to study the interaction between droplets in an emulsion are described. Various types of sterically stabilized emulsions are described: O/W emulsions stabilized with an A-B-A block copolymer of poly(ethylene oxide) (PEO, A) and poly(propylene oxide) (PPO, B); partially hydrolyzed poly(vinyl acetate) (PVAc); and W/O emulsions stabilized with an A-B-A block copolymer of poly(hydroxyl stearic acid) (PHS, A) and PEO (B).

On the basis of the above descriptions and details, it is clear that this book covers a wide range of topics: both fundamental and applied. It also highlights the engineering aspects of emulsion production and their characterization, both in the laboratory and during manufacture. It is hoped that this book will be of great help to emulsion research scientists, in both academia and industry.

I would like to thank the organizers – and in particular Dr Jean-Erik Poirier and Dr Alain Le Coroller – for giving me the opportunity to attend the Fifth World Congress and to edit this book.

October 2012 *Tharwat F. Tadros*

List of Contributors

Eugene V. Aksenenko
Ukrainian National Academy of
Sciences
Institute of Colloid Chemistry &
Chemistry of Water
42 Vernadsky Avenue
03680 Kyiv (Kiev)
Ukraine

Gabriela Calderó
Institute for Advanced Chemistry
of Catalonia
Consejo Superior de
Investigaciones Científicas
(IQAC-CSIC)
spain

and

CIBER en Bioingeniería,
Biomateriales y Nanomedicina
(CIBER-BBN)
Jordi Girona 18-26
08034 Barcelona
Spain

Michael Cooke
The University of Manchester
School of Chemical Engineering
and Analytical Sciences
Manchester
M60 1QD
UK

Jordi Esquena
Institute for Advanced Chemistry
of Catalonia
Consejo Superior de
Investigaciones Científicas
(IQAC-CSIC)
spain

and

CIBER de Bioingeniería
Biomateriales y Nanomedicina
(CIBER-BBN)
Jordi Girona 18-26
08034 Barcelona
Spain

Valentin B. Fainerman
Donetsk Medical University
Medical Physicochemical Centre
16 Ilych Avenue
83003 Donetsk
Ukraine

Steven Hall
University of Birmingham
School of Chemical Engineering
Edgbaston
Birmingham, B15 2TT
UK

Robin D. Hancocks
University of Birmingham
School of Chemical Engineering
Edgbaston
Birmingham, B15 2TT
UK

Aliyar Javadi
MPI of Colloids and Interfaces
Department Interfaces
Am Mühlenberg 1
14424 Potsdam-Golm
Germany

Isao Kobayashi
National Food Research Institute
National Agriculture and Food
Research Organization
Kannondai 2-1-12
Tsukuba, Ibaraki, 305-8642
Japan

Adam J. Kowalski
Process Science
Unilever R&D
Port Sunlight
Bebington
Wirral, CH63 3JW
UK

Jürgen Krägel
MPI of Colloids and Interfaces
Department Interfaces
Am Mühlenberg 1
14424 Potsdam-Golm
Germany

Martin E. Leser
Nestlé R&D Center
809 Collins Avenue
Marysville, OH 43040
U.S.A.

Reinhard Miller
MPI of Colloids and Interfaces
Department Interfaces
Am Mühlenberg 1
14424 Potsdam-Golm
Germany

Nenad Mucic
MPI of Colloids and Interfaces
Department Interfaces
Am Mühlenberg 1
14424 Potsdam-Golm
Germany

Mitsutoshi Nakajima
National Food Research Institute,
National Agriculture and Food
Research Organization
Kannondai 2-1-12
Tsukuba, Ibaraki, 305-8642
Japan

and

University of Tsukuba
Graduate School of Life and
Environmental Sciences 1-1-1
Tennoudai, Tsukuba
Ibaraki, 305-8572
Japan

Nima Niknafs
University of Birmingham
School of Chemical Engineering
Edgbaston
Birmingham, B15 2TT
UK

Ian T. Norton
University of Birmingham
School of Chemical Engineering
Edgbaston
Birmingham, B15 2TT
UK

Andrzej W. Pacek
University of Birmingham
School of Chemical Engineering
Edgbaston
Birmingham, B15 2TT
UK

Vincent Pradines
CNRS
Laboratoire de Chimie de
Coordination
UPR8241 205, route de Narbonne
31077 Toulouse Cedex 04
France

Altynay Sharipova
MPI of Colloids and Interfaces
Department Interfaces
Am Mühlenberg 1
14424 Potsdam-Golm
Germany

Conxita Solans
Institute for Advanced Chemistry
of Catalonia
Consejo Superior de
Investigaciones Científicas
(IQAC-CSIC)
spain

and

CIBER en Bioingeniería,
Biomateriales y Nanomedicina
(CIBER-BBN)
Jordi Girona 18-26
08034 Barcelona
Spain

Tharwat Tadros
89 Nash Grove Lane
Wokingham
Berkshire, RG40 4HE
UK

Jon Vilasau
Institute for Advanced Chemistry
of Catalonia
Consejo Superior de
Investigaciones Científicas
(IQAC-CSIC)
spain

and

CIBER en Bioingeniería,
Biomateriales y Nanomedicina
(CIBER-BBN)
Jordi Girona 18-26
08034 Barcelona
Spain

Goran T. Vladisavljević
Loughborough University
Chemical Engineering
Department
Ashby Road
Loughborough
Leicestershire
LE11 3TU
UK

1
Emulsion Formation, Stability, and Rheology
Tharwat F. Tadros

1.1
Introduction

Emulsions are a class of disperse systems consisting of two immiscible liquids [1–3]. The liquid droplets (the disperse phase) are dispersed in a liquid medium (the continuous phase). Several classes may be distinguished: oil-in-water (O/W), water-in-oil (W/O), and oil-in-oil (O/O). The latter class may be exemplified by an emulsion consisting of a polar oil (e.g., propylene glycol) dispersed in a nonpolar oil (paraffinic oil) and vice versa. To disperse two immiscible liquids, one needs a third component, namely, the emulsifier. The choice of the emulsifier is crucial in the formation of the emulsion and its long-term stability [1–3].

Emulsions may be classified according to the nature of the emulsifier or the structure of the system. This is illustrated in Table 1.1.

1.1.1
Nature of the Emulsifier

The simplest type is ions such as OH^- that can be specifically adsorbed on the emulsion droplet thus producing a charge. An electrical double layer can be produced, which provides electrostatic repulsion. This has been demonstrated with very dilute O/W emulsions by removing any acidity. Clearly that process is not practical. The most effective emulsifiers are nonionic surfactants that can be used to emulsify O/W or W/O. In addition, they can stabilize the emulsion against flocculation and coalescence. Ionic surfactants such as sodium dodecyl sulfate (SDS) can also be used as emulsifiers (for O/W), but the system is sensitive to the presence of electrolytes. Surfactant mixtures, for example, ionic and nonionic, or mixtures of nonionic surfactants can be more effective in emulsification and stabilization of the emulsion. Nonionic polymers, sometimes referred to as *polymeric surfactants*, for example, Pluronics, are more effective in stabilization of the emulsion, but they may suffer from the difficulty of emulsification (to produce small droplets) unless high energy is applied for the process. Polyelectrolytes such as poly(methacrylic

Emulsion Formation and Stability, First Edition. Edited by Tharwat F. Tadros.
© 2013 Wiley-VCH Verlag GmbH & Co. KGaA. Published 2013 by Wiley-VCH Verlag GmbH & Co. KGaA.

Table 1.1 Classification of emulsion types.

Nature of emulsifier	Structure of the system
Simple molecules and ions	Nature of internal and external phase: O/W, W/O
Nonionic surfactants	—
Surfactant mixtures	Micellar emulsions (microemulsions)
Ionic surfactants	Macroemulsions
Nonionic polymers	Bilayer droplets
Polyelectrolytes	Double and multiple emulsions
Mixed polymers and surfactants	Mixed emulsions
Liquid crystalline phases	—
Solid particles	—

acid) can also be applied as emulsifiers. Mixtures of polymers and surfactants are ideal in achieving ease of emulsification and stabilization of the emulsion. Lamellar liquid crystalline phases that can be produced using surfactant mixtures are very effective in emulsion stabilization. Solid particles that can accumulate at the O/W interface can also be used for emulsion stabilization. These are referred to as *Pickering emulsions*, whereby particles are made partially wetted by the oil phase and by the aqueous phase.

1.1.2
Structure of the System

1) **O/W and W/O macroemulsions**: These usually have a size range of 0.1–5 μm with an average of 1–2 μm.
2) **Nanoemulsions**: these usually have a size range of 20–100 nm. Similar to macroemulsions, they are only kinetically stable.
3) **Micellar emulsions or microemulsions**: these usually have the size range of 5–50 nm. They are thermodynamically stable.
4) **Double and multiple emulsions**: these are emulsions-of-emulsions, W/O/W, and O/W/O systems.
5) **Mixed emulsions**: these are systems consisting of two different disperse droplets that do not mix in a continuous medium. This chapter only deals with macroemulsions.

Several breakdown processes may occur on storage depending on particle size distribution and density difference between the droplets and the medium. Magnitude of the attractive versus repulsive forces determines flocculation. Solubility of the disperse droplets and the particle size distribution determine Ostwald ripening. Stability of the liquid film between the droplets determines coalescence. The other process is phase inversion.

1.1.3
Breakdown Processes in Emulsions

The various breakdown processes are illustrated in Figure 1.1. The physical phenomena involved in each breakdown process are not simple, and it requires analysis of the various surface forces involved. In addition, the above-mentioned processes may take place simultaneously rather than consecutively and this complicates the analysis. Model emulsions, with monodisperse droplets, cannot be easily produced, and hence, any theoretical treatment must take into account the effect of droplet size distribution. Theories that take into account the polydispersity of the system are complex, and in many cases, only numerical solutions are possible. In addition, measurements of surfactant and polymer adsorption in an emulsion are not easy and one has to extract such information from measurement at a planer interface.

In the following sections, a summary of each of the above-mentioned breakdown processes and details of each process and methods of its prevention are given.

1.1.4
Creaming and Sedimentation

This process results from external forces usually gravitational or centrifugal. When such forces exceed the thermal motion of the droplets (Brownain motion), a concentration gradient builds up in the system with the larger droplets moving faster to the top (if their density is lower than that of the medium) or to the bottom (if their density is larger than that of the medium) of the container. In the limiting cases, the droplets may form a close-packed (random or ordered) array at the top or bottom of the system with the remainder of the volume occupied by the continuous liquid phase.

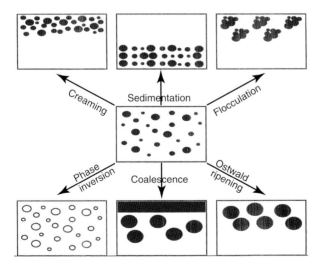

Figure 1.1 Schematic representation of the various breakdown processes in emulsions.

1.1.5
Flocculation

This process refers to aggregation of the droplets (without any change in primary droplet size) into larger units. It is the result of the van der Waals attraction that is universal with all disperse systems. Flocculation occurs when there is not sufficient repulsion to keep the droplets apart to distances where the van der Waals attraction is weak. Flocculation may be "strong" or "weak," depending on the magnitude of the attractive energy involved.

1.1.6
Ostwald Ripening (Disproportionation)

This results from the finite solubility of the liquid phases. Liquids that are referred to as *being immiscible* often have mutual solubilities that are not negligible. With emulsions, which are usually polydisperse, the smaller droplets will have larger solubility when compared with the larger ones (due to curvature effects). With time, the smaller droplets disappear and their molecules diffuse to the bulk and become deposited on the larger droplets. With time, the droplet size distribution shifts to larger values.

1.1.7
Coalescence

This refers to the process of thinning and disruption of the liquid film between the droplets with the result of fusion of two or more droplets into larger ones. The limiting case for coalescence is the complete separation of the emulsion into two distinct liquid phases. The driving force for coalescence is the surface or film fluctuations which results in close approach of the droplets whereby the van der Waals forces is strong thus preventing their separation.

1.1.8
Phase Inversion

This refers to the process whereby there will be an exchange between the disperse phase and the medium. For example, an O/W emulsion may with time or change of conditions invert to a W/O emulsion. In many cases, phase inversion passes through a transition state whereby multiple emulsions are produced.

1.2
Industrial Applications of Emulsions

Several industrial systems consist of emulsions of which the following is worth mentioning: food emulsion, for example, mayonnaise, salad creams, deserts, and beverages; personal care and cosmetics, for example, hand creams, lotions, hair sprays, and sunscreens; agrochemicals, for example, self-emulsifiable oils which produce emulsions on dilution with water, emulsion concentrates (EWs), and crop oil sprays; pharmaceuticals, for example, anesthetics of O/W emulsions, lipid emulsions, and double and multiple emulsions; and paints, for example, emulsions of alkyd resins and latex emulsions. Dry cleaning formulations – this may contain water droplets emulsified in the dry cleaning oil which is necessary to remove soils and clays. Bitumen emulsions: these are emulsions prepared stable in the containers, but when applied the road chippings, they must coalesce to form a uniform film of bitumen. Emulsions in the oil industry: many crude oils contain water droplets (for example, the North sea oil) and these must be removed by coalescence followed by separation. Oil slick dispersions: the oil spilled from tankers must be emulsified and then separated. Emulsification of unwanted oil: this is an important process for pollution control.

The above importance of emulsion in industry justifies a great deal of basic research to understand the origin of instability and methods to prevent their break down. Unfortunately, fundamental research on emulsions is not easy because model systems (e.g., with monodisperse droplets) are difficult to produce. In many cases, theories on emulsion stability are not exact and semiempirical approaches are used.

1.3
Physical Chemistry of Emulsion Systems

1.3.1
The Interface (Gibbs Dividing Line)

An interface between two bulk phases, for example, liquid and air (or liquid/vapor), or two immiscible liquids (oil/water) may be defined provided that a dividing line is introduced (Figure 1.2). The interfacial region is not a layer that is one-molecule thick. It is a region with thickness δ with properties different from the two bulk phases α and β.

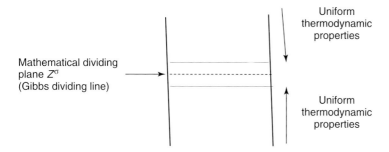

Figure 1.2 The Gibbs dividing line.

Using Gibbs model, it is possible to obtain a definition of the surface or interfacial tension γ.

The surface free energy dG^σ is made of three components: an entropy term $S^\sigma\,dT$, an interfacial energy term $A\,d\gamma$, and a composition term $\Sigma\,n_i d\mu_i$ (n_i is the number of moles of component i with chemical potential μ_i). The Gibbs–Deuhem equation is

$$dG^\sigma = -S^\sigma\,dT + A\,d\gamma + \sum n_i d\mu_i \qquad (1.1)$$

At constant temperature and composition

$$dG^\sigma = A\,d\gamma$$

$$\gamma = \left(\frac{\partial G^\sigma}{\partial A}\right)_{T,n_i} \qquad (1.2)$$

For a stable interface, γ is positive, that is, if the interfacial area increases G^σ increases. Note that γ is energy per unit area (mJ m^{-2}), which is dimensionally equivalent to force per unit length (mN m^{-1}), the unit usually used to define surface or interfacial tension.

For a curved interface, one should consider the effect of the radius of curvature. Fortunately, γ for a curved interface is estimated to be very close to that of a planer surface, unless the droplets are very small (<10 nm). Curved interfaces produce some other important physical phenomena that affect emulsion properties, for example, the Laplace pressure Δp, which is determined by the radii of curvature of the droplets

$$\Delta p = \gamma\left(\frac{1}{r_1} + \frac{1}{r_2}\right) \qquad (1.3)$$

where r_1 and r_2 are the two principal radii of curvature.

For a perfectly spherical droplet, $r_1 = r_2 = r$ and

$$\Delta p = \frac{2\gamma}{r} \qquad (1.4)$$

For a hydrocarbon droplet with radius 100 nm, and $\gamma = 50$ mN m^{-1}, $\Delta p = 10^6$ Pa (10 atm).

1.4
Thermodynamics of Emulsion Formation and Breakdown

Consider a system in which an oil is represented by a large drop 2 of area A_1 immersed in a liquid 2, which is now subdivided into a large number of smaller droplets with total area A_2 ($A_2 \gg A_1$) as shown in Figure 1.3. The interfacial tension γ_{12} is the same for the large and smaller droplets because the latter are generally in the region of 0.1 to few micrometers.

The change in free energy in going from state I to state II is made from two contributions: A surface energy term (that is positive) that is equal to $\Delta A \gamma_{12}$ (where $\Delta A = A_2 - A_1$). An entropy of dispersions term that is also positive

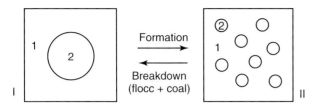

Figure 1.3 Schematic representation of emulsion formation and breakdown.

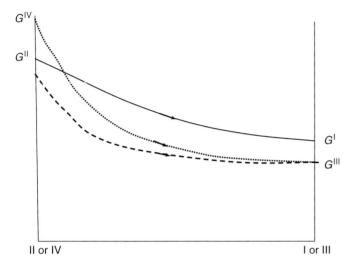

Figure 1.4 Free energy path in emulsion breakdown – (straight line) Flocc. + coal.; (dashed line) Flocc. + coal. + Sed.; and (dotted line) Flocc. + coal. + sed. + Ostwald ripening.

(since producing a large number of droplets is accompanied by an increase in configurational entropy), which is equal to $T\Delta S^{conf}$.

From the second law of thermodynamics

$$\Delta G^{form} = \Delta A \gamma_{12} - T\Delta S^{conf} \qquad (1.5)$$

In most cases, $\Delta A \gamma_{12} \gg -T\Delta S^{conf}$, which means that ΔG^{form} is positive, that is, the formation of emulsions is nonspontaneous and the system is thermodynamically unstable. In the absence of any stabilization mechanism, the emulsion will break by flocculation, coalescence, Ostwald ripening, or combination of all these processes. This is illustrated in Figure 1.4 that shows several paths for emulsion breakdown processes.

In the presence of a stabilizer (surfactant and/or polymer), an energy barrier is created between the droplets, and therefore, the reversal from state II to state I becomes noncontinuous as a result of the presence of these energy barriers. This is illustrated in Figure 1.5. In the presence of the above energy barriers, the system becomes kinetically stable.

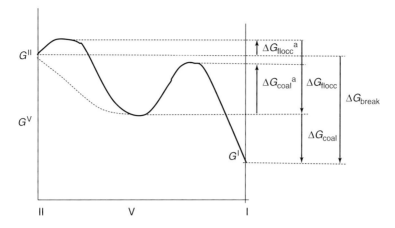

Figure 1.5 Schematic representation of free energy path for breakdown (flocculation and coalescence) for systems containing an energy barrier.

1.5
Interaction Energies (Forces) between Emulsion Droplets and Their Combinations

Generally speaking, there are three main interaction energies (forces) between emulsion droplets and these are discussed in the following sections.

1.5.1
van der Waals Attraction

The van der Waals attraction between atoms or molecules is of three different types: dipole–dipole (Keesom), dipole-induced dipole (Debye), and dispersion (London) interactions. The Keesom and Debye attraction forces are vectors, and although dipole–dipole or dipole-induced dipole attraction is large, they tend to cancel because of the different orientations of the dipoles. Thus, the most important are the London dispersion interactions that arise from charge fluctuations. With atoms or molecules consisting of a nucleus and electrons that are continuously rotating around the nucleus, a temporary dipole is created as a result of charge fluctuations. This temporary dipole induces another dipole in the adjacent atom or molecule. The interaction energy between two atoms or molecules G_a is short range and is inversely proportional to the sixth power of the separation distance r between the atoms or molecules

$$G_a = -\frac{\beta}{r^6} \tag{1.6}$$

where β is the London dispersion constant that is determined by the polarizability of the atom or molecule.

Hamaker [4] suggested that the London dispersion interactions between atoms or molecules in macroscopic bodies (such as emulsion droplets) can be added resulting in strong van der Waals attraction, particularly at close distances of

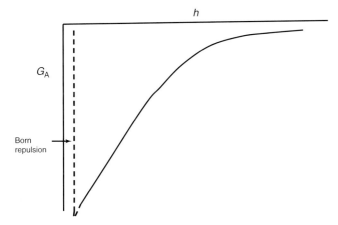

Figure 1.6 Variation of the van der Waals attraction energy with separation distance.

separation between the droplets. For two droplets with equal radii R, at a separation distance h, the van der Waals attraction G_A is given by the following equation (due to Hamaker)

$$G_A = -\frac{AR}{12h} \qquad (1.7)$$

where A is the effective Hamaker constant

$$A = \left(A_{11}^{1/2} - A_{22}^{1/2}\right)^2 \qquad (1.8)$$

where A_{11} and A_{22} are the Hamaker constants of droplets and dispersion medium, respectively.

The Hamaker constant of any material depends on the number of atoms or molecules per unit volume q and the London dispersion constant β

$$A = \pi^2 q^2 \beta \qquad (1.9)$$

G_A increases very rapidly with decrease of h (at close approach). This is illustrated in Figure 1.6 that shows the van der Waals energy–distance curve for two emulsion droplets with separation distance h.

In the absence of any repulsion, flocculation is very fast producing large clusters. To counteract the van der Waals attraction, it is necessary to create a repulsive force. Two main types of repulsion can be distinguished depending on the nature of the emulsifier used: electrostatic (due to the creation of double layers) and steric (due to the presence of adsorbed surfactant or polymer layers).

1.5.2
Electrostatic Repulsion

This can be produced by adsorption of an ionic surfactant as shown in Figure 1.7, which shows a schematic picture of the structure of the double layer according to

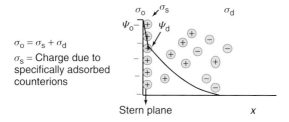

$\sigma_o = \sigma_s + \sigma_d$
σ_s = Charge due to specifically adsorbed counterions

Figure 1.7 Schematic representation of double layers produced by adsorption of an ionic surfactant.

Gouy–Chapman and Stern pictures [3]. The surface potential ψ_o decreases linearly to ψ_d (Stern or zeta potential) and then exponentially with increase of distance x. The double-layer extension depends on electrolyte concentration and valency (the lower the electrolyte concentration and the lower the valency the more extended the double layer is).

When charged colloidal particles in a dispersion approach each other such that the double layer begins to overlap (particle separation becomes less than twice the double-layer extension), repulsion occurs. The individual double layers can no longer develop unrestrictedly because the limited space does not allow complete potential decay [3, 4]. This is illustrated in Figure 1.8 for two flat, which clearly shows that when the separation distance h between the emulsion droplets becomes smaller than twice the double-layer extension, the potential at the midplane between the surfaces is not equal to zero (which would be the case when h is larger than twice the double-layer extension) plates.

The repulsive interaction G_{el} is given by the following expression:

$$G_{el} = 2\pi R \varepsilon_r \varepsilon_o \psi_o^2 \ln\left[1 + \exp(-\kappa h)\right] \tag{1.10}$$

where ε_r is the relative permittivity and ε_o is the permittivity of free space. κ is the Debye–Hückel parameter; $1/\kappa$ is the extension of the double layer (double-layer thickness) that is given by the expression

$$\left(\frac{1}{\kappa}\right) = \left(\frac{\varepsilon_r \varepsilon_o kT}{2 n_o Z_i^2 e^2}\right) \tag{1.11}$$

where k is the Boltzmann constant, T is the absolute temperature, n_o is the number of ions per unit volume of each type present in bulk solution, Z_i is the valency of the ions, and e is the electronic charge.

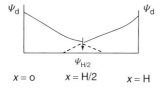

Figure 1.8 Schematic representation of double-layer overlap.

1.5 Interaction Energies (Forces) between Emulsion Droplets and Their Combinations

Values of $1/\kappa$ at various 1 : 1 electrolyte concentrations are given below

C (mol dm^{-3})	10^{-5}	10^{-4}	10^{-3}	10^{-2}	10^{-1}
$1/\kappa$ (nm)	100	33	10	3.3	1

The double-layer extension decreases with increase of electrolyte concentration. This means that the repulsion decreases with increase of electrolyte concentration as illustrated in Figure 1.9

$$G_T = G_{el} + G_A \tag{1.12}$$

A schematic representation of the force (energy)–distance curve according to the Derjaguin, Landau, Verwey, and Overbeek (DLVO) theory is given in Figure 1.10.

The above presentation is for a system at low electrolyte concentration. At large h, attraction prevails resulting in a shallow minimum (G_{sec}) of the order of few kilotesla units. At very short h, $V_A \gg G_{el}$, resulting in a deep primary minimum (several hundred kilotesla units). At intermediate h, $G_{el} > G_A$, resulting in a maximum (energy barrier) whose height depends on ψ_o (or ζ) and electrolyte concentration and valency – the energy maximum is usually kept >25 kT units.

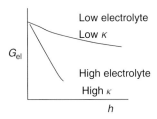

Figure 1.9 Variation of G_{el} with h at low and high electrolyte concentrations.

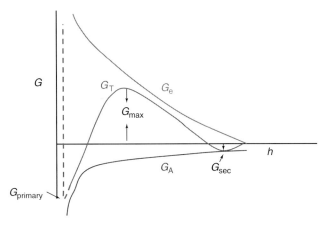

Figure 1.10 Total energy–distance curve according to the DLVO theory.

The energy maximum prevents close approach of the droplets, and flocculation into the primary minimum is prevented. The higher the value of ψ_o and the lower the electrolyte concentration and valency, the higher the energy maximum. At intermediate electrolyte concentrations, weak flocculation into the secondary minimum may occur.

Combination of van der Waals attraction and double-layer repulsion results in the well-known theory of colloid stability due to DLVO theory [5, 6].

1.5.3
Steric Repulsion

This is produced by using nonionic surfactants or polymers, for example, alcohol ethoxylates, or A-B-A block copolymers PEO-PPO-PEO (where PEO refers to polyethylene oxide and PPO refers to polypropylene oxide), as illustrated in Figure 1.11.

The "thick" hydrophilic chains (PEO in water) produce repulsion as a result of two main effects [7]:

1) Unfavorable mixing of the PEO chains, when these are in good solvent conditions (moderate electrolyte and low temperatures). This is referred to as the *osmotic* or *mixing free energy of interaction* that is given by the expression

$$\frac{G_{mix}}{kT} = \left(\frac{4\pi}{V_1}\right) \varphi_2^2 N_{av} \left(\frac{1}{2} - \chi\right) \left(\delta - \frac{h}{2}\right)^2 \left(3R + 2\delta + \frac{h}{2}\right) \quad (1.13)$$

V_1 is the molar volume of the solvent, ϕ_2 is the volume fraction of the polymer chain with a thickness δ, and χ is the Flory–Huggins interaction parameter. When $\chi < 0.5$, G_{mix} is positive and the interaction is repulsive. When $\chi > 0.5$, G_{mix} is negative and the interaction is attractive. When $\chi = 0.5$, $G_{mix} = 0$ and this is referred to as the θ-*condition*.

2) Entropic, volume restriction, or elastic interaction, G_{el}.
   This results from the loss in configurational entropy of the chains on significant overlap. Entropy loss is unfavorable and, therefore, G_{el} is always positive.

Combination of G_{mix}, G_{el} with G_A gives the total energy of interaction G_T (theory of steric stabilization)

$$G_T = G_{mix} + G_{el} + G_A \quad (1.14)$$

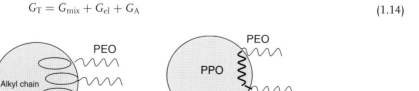

Figure 1.11 Schematic representation of adsorbed layers.

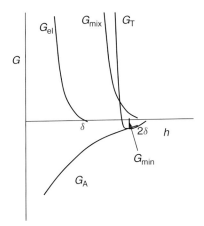

Figure 1.12 Schematic representation of the energy–distance curve for a sterically stabilized emulsion.

A schematic representation of the variation of G_{mix}, G_{el}, and G_A with h is given in Figure 1.12. G_{mix} increases very sharply with decrease of h when the latter becomes less than 2δ. G_{el} increases very sharply with decrease of h when the latter becomes smaller than δ. G_T increases very sharply with decrease of h when the latter becomes less than 2δ.

Figure 1.12 shows that there is only one minimum (G_{min}) whose depth depends on R, δ, and A. At a given droplet size and Hamaker constant, the larger the adsorbed layer thickness, the smaller the depth of the minimum. If G_{min} is made sufficiently small (large δ and small R), one may approach thermodynamic stability. This is illustrated in Figure 1.13 that shows the energy–distance curves as a function of δ/R. The larger the value of δ/R, the smaller the value of G_{min}. In this case, the system may approach thermodynamic stability as is the case with nanodispersions.

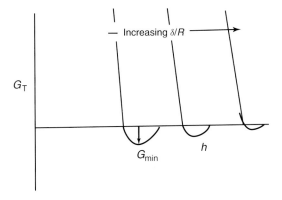

Figure 1.13 Variation of G_T with h at various δ/R values.

1.6
Adsorption of Surfactants at the Liquid/Liquid Interface

Surfactants accumulate at interfaces, a process described as adsorption. The simplest interfaces are the air/water (A/W) and O/W. The surfactant molecule orients itself at the interface with the hydrophobic portion orienting toward the hydrophobic phase (air or oil) and the hydrophilic portion orienting at the hydrophilic phase (water). This is schematically illustrated in Figure 1.14. As a result of adsorption, the surface tension of water is reduced from its value of 72 mN m^{-1} before adsorption to $\sim$30–40 mN m^{-1} and the interfacial tension for the O/W system decreases from its value of 50 mN m^{-1} (for an alkane oil) before adsorption to a value of 1–10 mN m^{-1} depending on the nature of the surfactant.

Two approaches can be applied to treat surfactant adsorption at the A/L and L/L interface [3]: Gibbs approach treats the process as an equilibrium phenomenon. In this case, one can apply the second law of thermodynamics. Equation of state approach whereby the surfactant film is treated as a "two-dimensional" layer with a surface pressure π. The Gibbs approach allows one to obtain the surfactant adsorption from surface tension measurements. The equation of state approach allows one to study the surfactant orientation at the interface. In this section, only the Gibbs approach is described.

1.6.1
The Gibbs Adsorption Isotherm

Gibbs derived a thermodynamic relationship between the variation of surface or interfacial tension with concentration and the amount of surfactant adsorbed Γ (moles per unit area), referred to as the *surface excess*. At equilibrium, the Gibbs free energy $dG^\sigma = 0$ and the Gibbs–Deuhem equation becomes

$$dG^\sigma = -S^\sigma dT + Ad\gamma + \sum n_i^\sigma d\mu_i = 0 \tag{1.15}$$

At constant temperature

$$Ad\gamma = -\sum n_i^\sigma d\mu_i \tag{1.16}$$

or

$$d\gamma = -\sum \frac{n_i^\sigma}{A} d\mu_i = -\sum \Gamma_i^\sigma d\mu_i \tag{1.17}$$

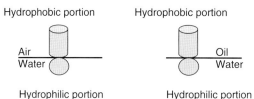

Figure 1.14 Schematic representation of orientation of surfactant molecules.

1.6 Adsorption of Surfactants at the Liquid/Liquid Interface

For a surfactant (component 2) adsorbed at the surface of a solvent (component 1)

$$-d\gamma = \Gamma_1^\sigma d\mu_1 + \Gamma_2^\sigma d\mu_2 \tag{1.18}$$

If the Gibbs dividing surface is used and the assumption $\Gamma_1^\sigma = 0$ is made

$$-d\gamma = \Gamma_{2,1}^\sigma d\mu_2 \tag{1.19}$$

The chemical potential of the surfactant μ_2 is given by the expression

$$\mu_2 = \mu_2^0 + RT \ln a_2^l \tag{1.20}$$

where μ_2^0 is the standard chemical potential and a_2^l is the activity of surfactant that is equal to $C_2 f_2 \sim x_2 f_2$ where C_2 is the concentration in moles per cubic decimeter and x_2 is the mole fraction that is equal to $C_2/(C_2 + 55.5)$ for a dilute solution, and f_2 is the activity coefficient that is also ~ 1 in dilute solutions.

Differentiating Eq. (1.20), one obtains

$$d\mu_2 = RT d\ln a_2^l \tag{1.21}$$

Combining Eqs. (1.19) and (1.21),

$$-d\gamma = \Gamma_{2,1}^\sigma RT d\ln a_2^l \tag{1.22}$$

or

$$\frac{d\gamma}{d\ln a_2^l} = -RT\Gamma_{2,1}^L \tag{1.23}$$

In dilute solutions, $f_2 \sim 1$ and

$$\frac{d\gamma}{d\ln C_2} = -\Gamma_2 RT \tag{1.24}$$

Equations (1.23) and (1.24) are referred to as the *Gibbs adsorption equations*, which show that Γ_2 can be determined from the experimental results of variation of γ with $\log C_2$ as illustrated in Figure 1.15 for the A/W and O/W interfaces.

Γ_2 can be calculated from the linear portion of the γ-$\log C$ curve just before the critical micelle concentration (cmc)

$$\text{Slope} = -\frac{d\gamma}{d\log C_2} = -2.303\Gamma_2 RT \tag{1.25}$$

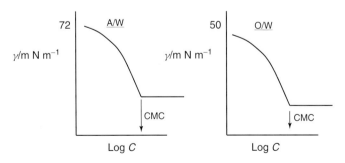

Figure 1.15 Surface or interfacial tension – log C curves.

From Γ_2, the area per molecule of surfactant (or ion) can be calculated

$$\text{Area/molecule} = \frac{1}{\Gamma_2 N_{av}}\,(m^2) = \frac{10^{18}}{\Gamma_2 N_{av}}\,(nm^2) \qquad (1.26)$$

N_{av} is Avogadro's constant that is equal to 6.023×10^{23}.

The area per surfactant ion or molecule gives information on the orientation of the ion or molecule at the interface. The area depends on whether the molecules lie flat or vertical at the interface. It also depends on the length of the alkyl chain length (if the molecules lie flat) or the cross-sectional area of the head group (if the molecules lie vertical. For example, for an ionic surfactant such as SDS, the area per molecule depends on the orientation. If the molecule lies flat, the area is determined by the area occupied by the alkyl chain and that by the sulfate head group. In this case, the area per molecule increases with increase in the alkyl chain length and will be in the range of $1-2\,nm^2$. In contrast, for vertical orientation, the area per molecule is determined by the cross-sectional area of the sulfate group, which is $\sim 0.4\,nm^2$ and virtually independent of the alkyl chain length. Addition of electrolytes screens the charge on the head group and hence the area per molecule decreases. For nonionic surfactants such as alcohol ethoxylates, the area per molecule for flat orientation is determined by the length of the alkyl chain and the number of ethylene oxide (EO) units. For vertical orientation, the area per molecule is determined by the cross-sectional area of the PEO chain and this increases with increase in the number of EO units.

At concentrations just before the break point, the slope of the γ-$\log C$ curve is constant

$$\left(\frac{\partial \gamma}{\partial \log C_2}\right) = \text{constant} \qquad (1.27)$$

This indicates that saturation of the interface occurs just below the cmc.

Above the break point ($C > $ cmc), the slope is 0,

$$\left(\frac{\partial \gamma}{\partial \log C_2}\right) = 0 \qquad (1.28)$$

or

$$\gamma = \text{constant} \times \log C_2 \qquad (1.29)$$

As γ remains constant above the cmc, then C_2 or a_2 of the monomer must remain constant.

Addition of surfactant molecules above the cmc must result in association to form micelles that have low activity, and hence, a_2 remains virtually constant.

The hydrophilic head group of the surfactant molecule can also affect its adsorption. These head groups can be unionized, for example, alcohol or PEO; weakly ionized, for example, COOH; or strongly ionized, for example, sulfates $-O-SO_3^-$, sulfonates $-SO_3^-$, or ammonium salts $-N^+(CH_3)_3^-$. The adsorption of the different surfactants at the A/W and O/W interface depends on the nature of the head group. With nonionic surfactants, repulsion between the head groups is smaller than with ionic head groups and adsorption occurs from dilute solutions; the cmc is low, typically 10^{-5} to 10^{-4} mol dm^{-3}. Nonionic surfactants with medium PEO form closely packed layers at C < cmc. Adsorption is slightly affected by moderate addition of electrolytes or change in the pH. Nonionic surfactant adsorption is relatively simple and can be described by the Gibbs adsorption equation.

With ionic surfactants, adsorption is more complicated depending on the repulsion between the head groups and addition of indifferent electrolyte. The Gibbs adsorption equation has to be solved to take into account the adsorption of the counterions and any indifferent electrolyte ions.

For a strong surfactant electrolyte such as $R-O-SO_3^-\ Na^+\ (R^-\ Na^+)$

$$\Gamma_2 = -\frac{1}{2RT}\left(\frac{\partial \gamma}{\partial \ln a\pm}\right) \quad (1.30)$$

The factor 2 in Eq. (1.30) arises because both surfactant ion and counterion must be adsorbed to maintain neutrality. $(\partial \gamma/\mathrm{dln}\ a\pm)$ is twice as large for an unionized surfactant molecule.

For a nonadsorbed electrolyte such as NaCl, any increase in $Na^+\cdot R^-$ concentration produces a negligible increase in Na^+ concentration ($d\mu_{Na}^+$ is negligible – $d\mu_{Cl}^-$ is also negligible.

$$\Gamma_2 = -\frac{1}{RT}\left(\frac{\partial \gamma}{\partial \ln C_{NaR}}\right) \quad (1.31)$$

which is identical to the case of nonionics.

The above analysis shows that many ionic surfactants may behave like nonionics in the presence of a large concentration of an indifferent electrolyte such as NaCl.

1.6.2
Mechanism of Emulsification

As mentioned before, to prepare an emulsions oil, water, surfactant, and energy are needed. This can be considered from a consideration of the energy required to expand the interface, $\Delta A \gamma$ (where ΔA is the increase in interfacial area when the bulk oil with area A_1 produces a large number of droplets with area A_2; $A_2 \gg A_1$, γ is the interfacial tension). As γ is positive, the energy to expand the interface is large and positive; this energy term cannot be compensated by the small entropy of dispersion $T\Delta S$ (which is also positive) and the total free energy of formation of an emulsion, ΔG given by Eq. (1.5) is positive. Thus, emulsion formation is nonspontaneous and energy is required to produce the droplets.

The formation of large droplets (few micrometers) as is the case for macroemulsions is fairly easy, and hence, high-speed stirrers such as the Ultraturrax or Silverson Mixer are sufficient to produce the emulsion. In contrast, the formation of small drops (submicrometer as is the case with nanoemulsions) is difficult and this requires a large amount of surfactant and/or energy. The high energy required for the formation of nanoemulsions can be understood from a consideration of the Laplace pressure Δp (the difference in pressure between inside and outside the droplet) as given by Eqs. (1.3) and (1.4).

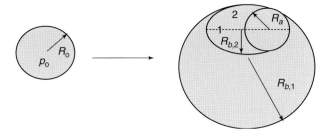

Figure 1.16 Illustration of increase in Laplace pressure when a spherical drop is deformed to a prolate ellipsoid.

To break up a drop into smaller ones, it must be strongly deformed and this deformation increases Δp. This is illustrated in Figure 1.16 that shows the situation when a spherical drop deforms into a prolate ellipsoid [8].

Near 1, there is only one radius of curvature R_a, whereas near 2, there are two radii of curvature $R_{b,1}$ and $R_{b,2}$. Consequently, the stress needed to deform the drop is higher for a smaller drop – as the stress is generally transmitted by the surrounding liquid via agitation, higher stresses need more vigorous agitation, and hence more energy is needed to produce smaller drops.

Surfactants play major roles in the formation of emulsions: by lowering the interfacial tension, p is reduced and hence the stress needed to break up a drop is reduced. Surfactants also prevent coalescence of newly formed drops.

Figure 1.17 shows an illustration of the various processes occurring during emulsification, break up of droplets, adsorption of surfactants, and droplet collision (which may or may not lead to coalescence) [8].

Each of the above processes occurs numerous times during emulsification and the timescale of each process is very short, typically a microsecond. This shows

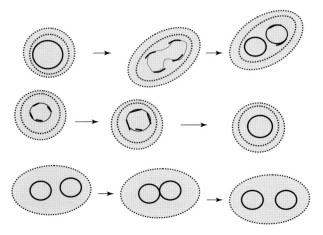

Figure 1.17 Schematic representation of the various processes occurring during emulsion formation. The drops are depicted by thin lines and the surfactant by heavy lines and dots.

that the emulsification process is a dynamic process and events that occur in a microsecond range could be very important.

To describe emulsion formation, one has to consider two main factors: hydrodynamics and interfacial science. In hydrodynamics, one has to consider the type of flow: laminar flow and turbulent flow. This depends on the Raynolds number as is discussed later.

To assess emulsion formation, one usually measures the droplet size distribution using, for example, laser diffraction techniques. A useful average diameter d is

$$d_{nm} = \left(\frac{S_m}{S_n}\right)^{1/(n-m)} \tag{1.32}$$

In most cases, d_{32} (the volume/surface average or Sauter mean) is used. The width of the size distribution can be given as the variation coefficient c_m, which is the standard deviation of the distribution weighted with d^m divided by the corresponding average d. Generally, C_2 is used that corresponds to d_{32}.

An alternative way to describe the emulsion quality is to use the specific surface area A (surface area of all emulsion droplets per unit volume of emulsion)

$$A = \pi s^2 = \frac{6\varphi}{d_{32}} \tag{1.33}$$

1.6.3
Methods of Emulsification

Several procedures may be applied for emulsion preparation, and these range from simple pipe flow (low agitation energy L); static mixers and general stirrers (low to medium energy, L–M); high-speed mixers such as the Ultraturrex (M); colloid mills and high-pressure homogenizers (high energy, H); and ultrasound generators (M–H). The method of preparation can be continuous (C) or batch-wise (B): pipe flow and static mixers – C; stirrers and Ultraturrax – B,C; colloid mill and high-pressure homogenizers – C; and ultrasound – B,C.

In all methods, there is liquid flow; unbounded and strongly confined flow. In the unbounded flow, any droplets are surrounded by a large amount of flowing liquid (the confining walls of the apparatus are far away from most of the droplets). The forces can be frictional (mostly viscous) or inertial. Viscous forces cause shear stresses to act on the interface between the droplets and the continuous phase (primarily in the direction of the interface). The shear stresses can be generated by laminar flow (LV) or turbulent flow (TV); this depends on the Reynolds number Re

$$Re = \frac{vl\rho}{\eta} \tag{1.34}$$

where v is the linear liquid velocity, ρ is the liquid density, and η is its viscosity. l is a characteristic length that is given by the diameter of flow through a cylindrical tube and by twice the slit width in a narrow slit.

For laminar flow, $Re \gtrsim 1000$, whereas for turbulent flow, $Re \gtrsim 2000$. Thus, whether the regime is linear or turbulent depends on the scale of the apparatus, the flow rate, and the liquid viscosity [9–12].

If the turbulent eddies are much larger than the droplets, they exert shear stresses on the droplets. If the turbulent eddies are much smaller than the droplets, inertial forces will cause disruption.

In bounded flow, other relations hold. If the smallest dimension of the part of the apparatus in which the droplets are disrupted (say a slit) is comparable to droplet size, other relations hold (the flow is always laminar). A different regime prevails if the droplets are directly injected through a narrow capillary into the continuous phase (injection regime), that is, membrane emulsification.

Within each regime, an essential variable is the intensity of the forces acting; the viscous stress during laminar flow $\sigma_{viscous}$ is given by

$$\sigma_{viscous} = \eta G \tag{1.35}$$

where G is the velocity gradient.

The intensity in turbulent flow is expressed by the power density ε (the amount of energy dissipated per unit volume per unit time); for laminar flow,

$$\varepsilon = \eta G^2 \tag{1.36}$$

The most important regimes are laminar/viscous (LV), turbulent/viscous (TV), and turbulent/inertial (TI). For water as the continuous phase, the regime is always TI. For higher viscosity of the continuous phase ($\eta_C = 0.1$ Pa s), the regime is TV. For still higher viscosity or a small apparatus (small l), the regime is LV. For very small apparatus (as is the case with most laboratory homogenizers), the regime is nearly always LV.

For the above regimes, a semiquantitative theory is available that can give the timescale and magnitude of the local stress σ_{ext}, the droplet diameter d, timescale of droplets deformation τ_{def}, timescale of surfactant adsorption τ_{ads}, and mutual collision of droplets.

An important parameter that describes droplet deformation is the Weber number We (which gives the ratio of the external stress over the Laplace pressure)

$$We = \frac{G\eta_C R}{2\gamma} \tag{1.37}$$

The viscosity of the oil plays an important role in the breakup of droplets; the higher the viscosity, the longer it will take to deform a drop. The deformation time τ_{def} is given by the ratio of oil viscosity to the external stress acting on the drop

$$\tau_{def} = \frac{\eta_D}{\sigma_{ext}} \tag{1.38}$$

The viscosity of the continuous phase η_C plays an important role in some regimes: for TI regime, η_C has no effect on droplet size. For turbulent viscous regime, larger η_C leads to smaller droplets. For laminar viscous, the effect is even stronger.

1.6.4
Role of Surfactants in Emulsion Formation

Surfactants lower the interfacial tension γ, and this causes a reduction in droplet size. The latter decrease with decrease in γ. For laminar flow, the droplet diameter is proportional to γ; for TI regime, the droplet diameter is proportional to $\gamma^{3/5}$.

The effect of reducing γ on the droplet size is illustrated in Figure 1.18, which shows a plot of the droplet surface area A and mean drop size d_{32} as a function of surfactant concentration m for various systems.

The amount of surfactant required to produce the smallest drop size will depend on its activity a (concentration) in the bulk that determines the reduction in γ, as given by the Gibbs adsorption equation

$$-d\gamma = RT\Gamma d\ln a \qquad (1.39)$$

where R is the gas constant, T is the absolute temperature, and Γ is the surface excess (number of moles adsorbed per unit area of the interface).

Γ increases with increase in surfactant concentration and eventually it reaches a plateau value (saturation adsorption). This is illustrated in Figure 1.19 for various emulsifiers.

The value of γ obtained depends on the nature of the oil and surfactant used; small molecules such as nonionic surfactants lower γ more than polymeric surfactants such as PVA.

Another important role of the surfactant is its effect on the interfacial dilational modulus ε

$$\varepsilon = \frac{d\gamma}{d\ln A} \qquad (1.40)$$

During emulsification, an increase in the interfacial area A takes place and this causes a reduction in Γ. The equilibrium is restored by adsorption of surfactant from the bulk, but this takes time (shorter times occur at higher surfactant activity).

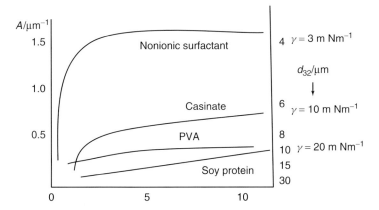

Figure 1.18 Variation of A and d_{32} with m for various surfactant systems.

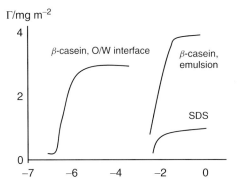

Figure 1.19 Variation of Γ (mg m^{-2}) with log C_{eq}/wt% – the oils are β-casein (O–W interface) toluene, β-casein (emulsions) soybean, and SDS benzene.

Thus, ε is small at small a and also at large a. Because of the lack or slowness of equilibrium with polymeric surfactants, ε will not be the same for expansion and compression of the interface.

In practice, surfactant mixtures are used and these have pronounced effects on γ and ε. Some specific surfactant mixtures give lower γ values than either of the two individual components. The presence of more than one surfactant molecule at the interface tends to increase ε at high surfactant concentrations. The various components vary in surface activity. Those with the lowest γ tend to predominate at the interface, but if present at low concentrations, it may take long time before reaching the lowest value. Polymer–surfactant mixtures may show some synergetic surface activity.

1.6.5
Role of Surfactants in Droplet Deformation

Apart from their effect on reducing γ, surfactants play major roles in deformation and breakup of droplets – this is summarized as follows. Surfactants allow the existence of interfacial tension gradients, which is crucial for the formation of stable droplets. In the absence of surfactants (clean interface), the interface cannot withstand a tangential stress; the liquid motion will be continuous (Figure 1.20a).

If a liquid flows along the interface with surfactants, the latter will be swept downstream causing an interfacial tension gradient (Figure 1.20b). A balance of forces will be established

$$\eta \left[\frac{dV_x}{dy} \right]_{y=0} = -\frac{d\gamma}{dx} \quad (1.41)$$

If the γ-gradient can become large enough, it will arrest the interface. If the surfactant is applied at one site of the interface, a γ-gradient is formed that will cause the interface to move roughly at a velocity given by

$$v = 1.2 \, [\eta \rho z]^{-1/3} \, |\Delta \gamma|^{2/3} \quad (1.42)$$

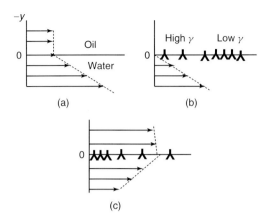

Figure 1.20 Interfacial tension gradients and flow near an oil/water interface: (a) no surfactant; (b) velocity gradient causes an interfacial tension gradient; and (c) interfacial tension gradient causes flow (Marangoni effect).

The interface will then drag some of the bordering liquid with it (Figure 1.20c).

Interfacial tension gradients are very important in stabilizing the thin liquid film between the droplets, which is very important during the beginning of emulsification (films of the continuous phase may be drawn through the disperse phase and collision is very large). The magnitude of the γ-gradients and the Marangoni effect depends on the surface dilational modulus ε, which for a plane interface with one surfactant-containing phase, is given by the expression

$$\varepsilon = \frac{-d\gamma/d\ln\Gamma}{(1 + 2\xi + 2\xi^2)^{1/2}} \tag{1.43}$$

$$\xi = \frac{dm_C}{d\Gamma}\left(\frac{D}{2\omega}\right)^{1/2} \tag{1.44}$$

$$\omega = \frac{d\ln A}{dt} \tag{1.45}$$

where D is the diffusion coefficient of the surfactant and ω represents a timescale (time needed for doubling the surface area) that is roughly equal to τ_{def}.

During emulsification, ε is dominated by the magnitude of the denominator in Eq. (1.43) because ζ remains small. The value of $dm_C/d\Gamma$ tends to go to very high values when Γ reaches its plateau value; ε goes to a maximum when m_C is increased.

For conditions that prevail during emulsification, ε increases with m_C and it is given by the relationship

$$\varepsilon = \frac{d\pi}{d\ln\Gamma} \tag{1.46}$$

where π is the surface pressure ($\pi = \gamma_o - \gamma$). Figure 1.21 shows the variation of π with $\ln\Gamma$; ε is given by the slope of the line.

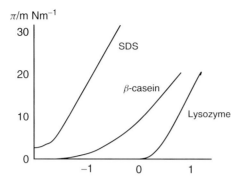

Figure 1.21 π versus $\ln \Gamma$ for various emulsifiers.

The SDS shows a much higher ε value when compared with β-casein and lysozyme – this is because the value of Γ is higher for SDS. The two proteins show difference in their ε values, which may be attributed to the conformational change that occurs on adsorption.

The presence of a surfactant means that during emulsification the interfacial tension needs not to be the same every where (Figure 1.20). This has two consequences: (i) the equilibrium shape of the drop is affected and (ii) any γ-gradient formed will slow down the motion of the liquid inside the drop (this diminishes the amount of energy needed to deform and break up the drop).

Another important role of the emulsifier is to prevent coalescence during emulsification. This is certainly not due to the strong repulsion between the droplets, because the pressure at which two drops are pressed together is much greater than the repulsive stresses. The counteracting stress must be due to the formation of γ-gradients. When two drops are pushed together, liquid will flow out from the thin layer between them, and the flow will induce a γ-gradient. This was shown in Figure 1.20c. This produces a counteracting stress given by

$$\tau_{\Delta\gamma} \approx \frac{2|\Delta\gamma|}{(1/2)d} \tag{1.47}$$

The factor 2 follows from the fact that two interfaces are involved. Taking a value of $\Delta\gamma = 10$ mN m^{-1}, the stress amounts to 40 kPa (which is of the same order of magnitude as the external stress).

Closely related to the above mechanism is the Gibbs-Marangoni effect [13–17], schematically represented in Figure 1.22. The depletion of surfactant in the thin film between approaching drops results in γ-gradient without liquid flow being involved. This results in an inward flow of liquid that tends to drive the drops apart.

The Gibbs–Marangoni effect also explains the Bancroft rule, which states that the phase in which the surfactant is most soluble forms the continuous phase. If the surfactant is in the droplets, a γ-gradient cannot develop and the drops would

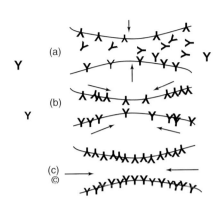

Figure 1.22 Schematic representation of the Gibbs–Marangoni effect for two approaching drops.

be prone to coalescence. Thus, surfactants with HLB > 7 (hydrophilic–lipophilic balance) tend to form O/W emulsions and HLB < 7 tend to form W/O emulsions.

The Gibbs–Marangoni effect also explains the difference between surfactants and polymers for emulsification – polymers give larger drops when compared with surfactants. Polymers give a smaller value of ε at small concentrations when compared to surfactants (Figure 1.21).

Various other factors should also be considered for emulsification: The disperse phase volume fraction ϕ. An increase in ϕ leads to increase in droplet collision and hence coalescence during emulsification. With increase in ϕ, the viscosity of the emulsion increases and could change the flow from being turbulent to being laminar (LV regime).

The presence of many particles results in a local increase in velocity gradients. This means that G increases. In turbulent flow, increase in ϕ will induce turbulence depression. This results in larger droplets. Turbulence depression by added polymers tends to remove the small eddies, resulting in the formation of larger droplets.

If the mass ratio of surfactant to continuous phase is kept constant, increase in ϕ results in decrease in surfactant concentration and hence an increase in γ_{eq} resulting in larger droplets. If the mass ratio of surfactant to disperse phase is kept constant, the above changes are reversed.

General conclusions cannot be drawn because several of the above-mentioned mechanism may come into play. Experiments using a high-pressure homogenizer at various ϕ values at constant initial m_C (regime TI changing to TV at higher ϕ) showed that with increasing ϕ (>0.1) the resulting droplet diameter increased and the dependence on energy consumption became weaker. Figure 1.23 shows a comparison of the average droplet diameter versus power consumption using different emulsifying machines. It can be seen that the smallest droplet diameters were obtained when using the high-pressure homogenizers.

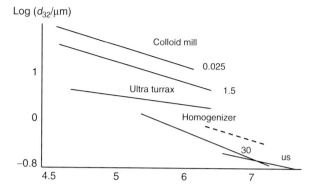

Figure 1.23 Average droplet diameters obtained in various emulsifying machines as a function of energy consumption p – the number near the curves denotes the viscosity ratio λ – the results for the homogenizer are for $\phi = 0.04$ (solid line) and $\phi = 0.3$ (broken line) – us means ultrasonic generator.

1.7
Selection of Emulsifiers

1.7.1
The Hydrophilic–Lipophilic Balance (HLB) Concept

The selection of different surfactants in the preparation of either O/W or W/O emulsions is often still made on an empirical basis. A semiempirical scale for selecting surfactants is the HLB number developed by Griffin [18]. This scale is based on the relative percentage of hydrophilic to lipophilic (hydrophobic) groups in the surfactant molecule(s). For an O/W emulsion droplet, the hydrophobic chain resides in the oil phase, whereas the hydrophilic head group resides in the aqueous phase. For a W/O emulsion droplet, the hydrophilic group(s) reside in the water droplet, whereas the lipophilic groups reside in the hydrocarbon phase.

Table 1.2 gives a guide to the selection of surfactants for a particular application. The HLB number depends on the nature of the oil. As an illustration, Table 1.3 gives the required HLB numbers to emulsify various oils.

Table 1.2 Summary of HLB ranges and their applications.

HLB range	Application
3–6	W/O emulsifier
7–9	Wetting agent
8–18	O/W emulsifier
13–15	Detergent
15–18	Solubilizer

Table 1.3 Required HLB numbers to emulsify various oils.

Oil	W/O emulsion	O/W emulsion
Paraffin oil	4	10
Beeswax	5	9
Linolin, anhydrous	8	12
Cyclohexane	—	15
Toluene	—	15

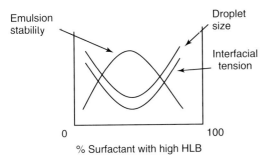

Figure 1.24 Variation of emulsion stability, droplet size, and interfacial tension with percentage surfactant with high HLB number.

The relative importance of the hydrophilic and lipophilic groups was first recognized when using mixtures of surfactants containing varying proportions of a low and high HLB number.

The efficiency of any combination (as judged by phase separation) was found to pass a maximum when the blend contained a particular proportion of the surfactant with the higher HLB number. This is illustrated in Figure 1.24 that shows the variation of emulsion stability, droplet size, and interfacial tension with percentage surfactant with high HLB number.

The average HLB number may be calculated from additivity

$$\text{HLB} = x_1 \text{HLB}_1 + x_2 \text{HLB}_2 \tag{1.48}$$

where x_1 and x_2 are the weigh fractions of the two surfactants with HLB_1 and HLB_2.

Griffin developed simple equations for calculation of the HLB number of relatively simple nonionic surfactants. For a polyhydroxy fatty acid ester

$$\text{HLB} = 20\left(1 - \frac{S}{A}\right) \tag{1.49}$$

S is the saponification number of the ester and A is the acid number. For a glyceryl monostearate, $S = 161$ and $A = 198$; the HLB is 3.8 (suitable for W/O emulsion).

For a simple alcohol ethoxylate, the HLB number can be calculated from the weight percentage of ethylene oxide (E) and polyhydric alcohol (P)

$$\text{HLB} = \frac{E + P}{5} \tag{1.50}$$

If the surfactant contains PEO as the only hydrophilic group, the contribution from one OH group neglected

$$\text{HLB} = \frac{E}{5} \tag{1.51}$$

For a nonionic surfactant $C_{12}H_{25}-O-(CH_2-CH_2-O)_6$, the HLB is 12 (suitable for O/W emulsion).

The above simple equations cannot be used for surfactants containing propylene oxide or butylene oxide. They cannot be also applied for ionic surfactants. Davies [19, 20] devised a method for calculating the HLB number for surfactants from their chemical formulae, using empirically determined group numbers. A group number is assigned to various component groups. A summary of the group numbers for some surfactants is given in Table 1.4.

The HLB is given by the following empirical equation:

$$\text{HLB} = 7 + \sum(\text{hydrophilic group Nos}) - \sum(\text{lipohilic group Nos}) \tag{1.52}$$

Davies has shown that the agreement between HLB numbers calculated from the above equation and those determined experimentally is quite satisfactory.

Various other procedures were developed to obtain a rough estimate of the HLB number. Griffin found good correlation between the cloud point of 5% solution of various ethoxylated surfactants and their HLB number.

Davies [17, 18] attempted to relate the HLB values to the selective coalescence rates of emulsions. Such correlations were not realized since it was found that the emulsion stability and even its type depends to a large extent on the method of

Table 1.4 HLB group numbers.

Hydrophilic	Group number
$-SO_4Na^+$	38.7
$-COO$	21.2
$-COONa$	19.1
N(tertiary amine)	9.4
Ester (sorbitan ring)	6.8
$-O-$	1.3
CH–(sorbitan ring)	0.5
Lipophilic	
$(-CH-), (-CH_2-), CH_3$	0.475
Derived	
$-CH_2-CH_2-O$	0.33
$-CH_2-CH_2-CH_2-O-$	−0.15

dispersing the oil into the water and vice versa. At best, the HLB number can only be used as a guide for selecting optimum compositions of emulsifying agents.

One may take any pair of emulsifying agents, which fall at opposite ends of the HLB scale, for example, Tween 80 (sorbitan monooleate with 20 mol EO, HLB = 15) and Span 80 (sorbitan monooleate, HLB = 5) using them in various proportions to cover a wide range of HLB numbers. The emulsions should be prepared in the same way, with a small percentage of the emulsifying blend. The stability of the emulsions is then assessed at each HLB number from the rate of coalescence or qualitatively by measuring the rate of oil separation. In this way, one may be able to find the optimum HLB number for a given oil. Having found the most effective HLB value, various other surfactant pairs are compared at this HLB value to find the most effective pair.

1.7.2
The Phase Inversion Temperature (PIT) Concept

Shinoda and coworkers [21, 22] found that many O/W emulsions stabilized with nonionic surfactants undergo a process of inversion at a critical temperature (phase inversion temperature, PIT). The PIT can be determined by following the emulsion conductivity (small amount of electrolyte is added to increase the sensitivity) as a function of temperature. The conductivity of the O/W emulsion increases with increase of temperature till the PIT is reached, above which there will be a rapid reduction in conductivity (W/O emulsion is formed). Shinoda and coworkers found that the PIT is influenced by the HLB number of the surfactant. The size of the emulsion droplets was found to depend on the temperature and HLB number of the emulsifiers. The droplets are less stable toward coalescence close to the PIT. However, by rapid cooling of the emulsion, a stable system may be produced. Relatively stable O/W emulsions were obtained when the PIT of the system was 20–65 °C higher than the storage temperature. Emulsions prepared at a temperature just below the PIT followed by rapid cooling generally have smaller droplet sizes. This can be understood if one considers the change of interfacial tension with temperature as is illustrated in Figure 1.25. The interfacial tension decreases with increase of temperature reaching a minimum close to the PIT, after which it increases.

Thus, the droplets prepared close to the PIT are smaller than those prepared at lower temperatures. These droplets are relatively unstable toward coalescence near the PIT, but by rapid cooling of the emulsion, one can retain the smaller size. This procedure may be applied to prepare mini (nano) emulsions.

The optimum stability of the emulsion was found to be relatively insensitive to changes in the HLB value or the PIT of the emulsifier, but instability was very sensitive to the PIT of the system.

It is essential, therefore, to measure the PIT of the emulsion as a whole (with all other ingredients).

At a given HLB value, stability of the emulsions against coalescence increases markedly as the molar mass of both the hydrophilic and lipophilic components

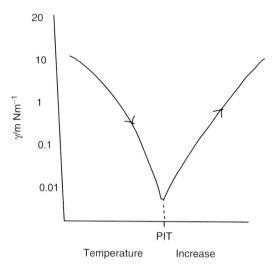

Figure 1.25 Variation of interfacial tension with temperature increase for an O/W emulsion.

increases. The enhanced stability using high-molecular-weight surfactants (polymeric surfactants) can be understood from a consideration of the steric repulsion that produces more stable films. Films produced using macromolecular surfactants resist thinning and disruption thus reducing the possibility of coalescence. The emulsions showed maximum stability when the distribution of the PEO chains was broad. The cloud point is lower but the PIT is higher than in the corresponding case for narrow size distributions. The PIT and HLB number are directly related parameters.

Addition of electrolytes reduces the PIT and hence an emulsifier with a higher PIT value is required when preparing emulsions in the presence of electrolytes. Electrolytes cause dehydration of the PEO chains and in effect this reduces the cloud point of the nonionic surfactant. One needs to compensate for this effect by using a surfactant with higher HLB. The optimum PIT of the emulsifier is fixed if the storage temperature is fixed.

In view of the above correlation between PIT and HLB and the possible dependence of the kinetics of droplet coalescence on the HLB number, Sherman and coworkers suggested the use of PIT measurements as a rapid method for assessing emulsion stability. However, one should be careful in using such methods for assessment of the long-term stability because the correlations were based on a very limited number of surfactants and oils.

Measurement of the PIT can at best be used as a guide for the preparation of stable emulsions. Assessment of the stability should be evaluated by following the droplet size distribution as a function of time using a Coulter counter or light diffraction techniques. Following the rheology of the emulsion as a function of time and temperature may also be used for assessment of the stability against coalescence. Care should be taken in analyzing the rheological results. Coalescence results in an increase in the droplet size and this is usually followed by a reduction in the viscosity of the emulsion. This trend is only observed if the coalescence is not accompanied by flocculation of the emulsion droplets (which results in an increase in the viscosity). Ostwald ripening can also complicate the analysis of the rheological data.

1.7.3
The Cohesive Energy Ratio (CER) Concept

Beerbower and Hills [23] considered the dispersing tendency on the oil and water interfaces of the surfactant or emulsifier in terms of the ratio of the cohesive energies of the mixtures of oil with the lipophilic portion of the surfactant and the water with the hydrophilic portion. They used the Winsor R_o concept, which is the ratio of the intermolecular attraction of oil molecules (O) and lipophilic portion of surfactant (L), C_{LO}, to that of water(W) and hydrophilic portion (H), C_{HW}

$$R_o = \frac{C_{LO}}{C_{HW}} \tag{1.53}$$

Several interaction parameters may be identified at the oil and water sides of the interface. One can identify at least nine interaction parameters as schematically represented in Figure 1.26.

In the absence of emulsifier, there will be only three interaction parameters: C_{OO}, C_{WW}, C_{OW}; if $C_{OW} \ll C_{WW}$, the emulsion breaks.

The above interaction parameters may be related to the Hildebrand solubility parameter [24] δ (at the oil side of the interface) and the Hansen [25] nonpolar, hydrogen bonding, and polar contributions to δ at the water side of the interface. The solubility parameter of any component is related to its heat of vaporization ΔH by the expression

$$\delta^2 = \frac{\Delta H - RT}{V_m} \tag{1.54}$$

where V_m is the molar volume.

Hansen considered δ (at the water side of the interface) to consist of three main contributions, a dispersion contribution, δ_d; a polar contribution, δ_p; and a

C_{LL}, C_{OO}, C_{LO} (at oil side)
C_{HH}, C_{WW}, C_{HW} (at water side)
C_{LW}, C_{HO}, C_{LH} (at the interface)

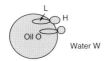

Figure 1.26 The cohesive energy ratio concept.

hydrogen-bonding contribution, δ_h. These contributions have different weighting factors

$$\delta^2 = \delta_d^2 + 0.25\delta_p^2 + \delta_h^2 \tag{1.55}$$

Beerbower and Hills used the following expression for the HLB number

$$\text{HLB} = 20\left(\frac{M_H}{M_L + M_H}\right) = 20\left(\frac{V_H \rho_H}{V_L \rho_L + V_H \rho_H}\right) \tag{1.56}$$

where M_H and M_L are the molecular weights of the hydrophilic and lipophilic portions of the surfactants, respectively. V_L and V_H are their corresponding molar volumes whereas ρ_H and ρ_L are the densities, respectively.

The cohesive energy ratio (CER) was originally defined by Winsor, Eq. (1.53).

When $C_{LO} > C_{HW}$, $R > 1$ and a W/O emulsion forms. If $C_{LO} < C_{HW}$, $R < 1$ and an O/W emulsion forms. If $C_{LO} = C_{HW}$, $R = 1$ and a planer system results – this denotes the inversion point.

R_o can be related to V_L, δ_L and V_H, δ_H by the expression

$$R_o = \frac{V_L \delta_L^2}{V_H \delta_H^2} \tag{1.57}$$

Using Eq. (1.55)

$$R_o = \frac{V_L \left(\delta_d^2 + 0.25\delta_p^2 + 0.25\delta_h^2\right)_L}{V_h \left(\delta_d^2 + 0.25\delta_p^2 + 0.25\delta_h^2\right)_H} \tag{1.58}$$

Combining Eqs. (1.57) and (1.58), one obtains the following general expression for the CER

$$R_o = \left(\frac{20}{\text{HLB}} - 1\right) \frac{\rho_h \left(\delta_d^2 + 025\delta_p^2 + 0.25\delta_h^2\right)_L}{\rho_L \left(\delta_d^2 + 0.25\delta_p^2 + 0.25\delta_p^2\right)_L} \tag{1.59}$$

For O/W systems, HLB = 12–15 and R_o = 0.58–0.29 ($R_o < 1$). For W/O systems, HLB = 5–6 and R_o = 2.3–1.9 ($R_o > 1$). For a planer system, HLB = 8–10 and R_o = 1.25–0.85 ($R_o \sim 1$)

The R_o equation combines both the HLB and cohesive energy densities – it gives a more quantitative estimate of emulsifier selection. R_o considers HLB, molar volume, and chemical match. The success of this approach depends on the availability of data on the solubility parameters of the various surfactant portions. Some values are tabulated in the book by Barton [26].

1.7.4
The Critical Packing Parameter (CPP) for Emulsion Selection

The critical packing parameter (CPP) is a geometric expression relating the hydrocarbon chain volume (v) and length (l) and the interfacial area occupied by the head group (a) [27]

1.7 Selection of Emulsifiers

$$\text{CPP} = \frac{v}{l_c a_o} \tag{1.60}$$

a_o is the optimal surface area per head group and l_c is the critical chain length.

Regardless of the shape of any aggregated structure (spherical or cylindrical micelle or a bilayer), no point within the structure can be farther from the hydrocarbon-water surface than l_c. The critical chain length, l_c, is roughly equal but less than the fully extended length of the alkyl chain.

The above concept can be applied to predict the shape of an aggregated structure. Consider a spherical micelle with radius r and aggregation number n; the volume of the micelle is given by

$$\left(\frac{4}{3}\right)\pi r^3 = nv \tag{1.61}$$

where v is the volume of a surfactant molecule.

The area of the micelle is given by

$$4\pi r^2 = na_o \tag{1.62}$$

where a_o is the area per surfactant head group.

Combining Eqs. (1.61) and (1.62)

$$a_o = \frac{3v}{r} \tag{1.63}$$

The cross-sectional area of the hydrocarbon chain a is given by the ratio of its volume to its extended length l_c

$$a = \frac{v}{l_c} \tag{1.64}$$

From Eqs. (1.63) and (1.64)

$$\text{CPP} = \frac{a}{a_o} = \left(\frac{1}{3}\right)\left(\frac{r}{l_c}\right) \tag{1.65}$$

As $r < l_c$, then CPP $\leq (1/3)$.

For a cylindrical micelle with length d and radius r

$$\text{Volume of the micelle} = \pi r^2 d = nv \tag{1.66}$$

$$\text{Area of the micelle} = 2\pi rd = na_o \tag{1.67}$$

Combining Eqs. (1.66) and (1.67)

$$a_o = \frac{2v}{r} \tag{1.68}$$

$$a = \frac{v}{l_c} \tag{1.69}$$

$$\text{CPP} = \frac{a}{a_o} = \left(\frac{1}{2}\right)\left(\frac{r}{l_c}\right) \tag{1.70}$$

As $r < l_c$, then $(1/3) < \text{CPP} \leq (1/2)$.

For vesicles (liposomes), $1 > \text{CPP} \geq (2/3)$ and for lamellar micelles $P \sim 1$. For inverse micelles, CPP > 1. A summary of the various shapes of micelles and their CPP is given in Table 1.5.

Table 1.5 Critical packing parameter and various shapes of micelles.

Lipid	Critical packing parameter v/anlc	Critical packing shape	Structures formed
Single-chained lipids (surfactants) with large head group areas: SDS in low salt	<1/3	Cone	Spherical micelles
Single-chained lipids with small head group areas: SDS and CTAB in high salt Nonionic lipids	1/3–1/2	Truncated cone	Cylindrical micelles
Double-chained lipids with large head group areas, fluid chains: Phosphatidyl choline (lecithin) Phosphatidyl serine Phosphatidyl glycerol Phosphatidyl inositol Phosphatidic acid Sphingomyelin, DGDG[a] Dihexadecyl phosphate Dialkyl dimethyl ammonium Salts	1/2–1	Truncated cone	Flexible bilayers, vesicles
Double-chained lipids with small head group areas, anionic lipids in high salt, saturated frozen chains: Phosphatidyl ethanolamine Phosphatidyl serine + Ca^{2+}	~1	Cylinder	Planar bilayers

Table 1.5 (continued)

Lipid	Critical packing parameter v/anlc	Critical packing shape	Structures formed
Double-chained lipids with small head group areas, nonionic lipids, poly(cis) unsaturated chains, high T: Unsaturated, phosphatidyl ethanolamine Cardiolipin} + Ca^{2+} Phosphatidic acid + Ca^{2+} Cholesterol, MGDG^b	>1	Inverted truncated cone or wedge	Inverted miscelles

^{a}DGDG, digalactosyl diglyceride, diglucosyldiglyceride;
^{b}MGDG, monogalactosyl diglyceride, monoglucosyl diglyceride.

Surfactants that make spherical micelles with the above packing constraints, that is, CPP $\leq (1/3)$, are more suitable for O/W emulsions. Surfactants with CPP > 1, that is, forming inverted micelles, are suitable for the formation of W/O emulsions.

1.8
Creaming or Sedimentation of Emulsions

This is the result of gravity, when the density of the droplets and the medium are not equal. When the density of the disperse phase is lower than that of the medium, creaming occurs, whereas if the density of the disperse phase is higher than that of the medium, sedimentation occurs. Figure 1.27 gives a schematic picture for creaming of emulsions for three cases [28].

Case (a) represents the situation for small droplets (< 0.1 μm, i.e., nanoemulsions) whereby the Brownian diffusion kT (where k is the Boltzmann constant and T is the absolute temperature) exceeds the force of gravity (mass × acceleration due to gravity g)

$$kT \gg \frac{4}{3}\pi R^3 \Delta\rho g L \tag{1.71}$$

where R is the droplet radius, $\Delta\rho$ is the density difference between the droplets and the medium, and L is the height of the container.

Case (b) represents emulsions consisting of "monodisperse" droplets with radius > 1 μm. In this case, the emulsion separates into two distinct layers with the droplets forming a cream or sediment leaving the clear supernatant liquid. This situation is seldom observed in practice. Case (c) is that for a polydisperse

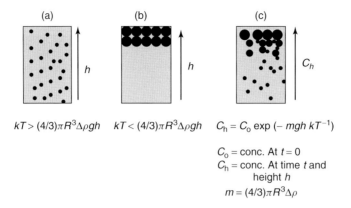

Figure 1.27 (a–c) Representation of creaming of emulsions.

(practical) emulsions, in which case the droplets will cream or sediment at various rates. In the last case, a concentration gradient build up with the larger droplets staying at the top of the cream layer or the bottom of the sediment

$$C(h) = C_o \exp\left(-\frac{mgh}{kT}\right) \tag{1.72}$$

$C(h)$ is the concentration (or volume fraction ϕ) of droplets at height h, whereas C_o is the concentration at the top or bottom of the container.

1.8.1
Creaming or Sedimentation Rates

1) Very dilute emulsions ($\phi < 0.01$). In this case, the rate could be calculated using Stokes' law that balances the hydrodynamic force with gravity force

$$\text{Hydrodynamic force} = 6\pi \eta_o R v_o \tag{1.73}$$

$$\text{Gravity force} = \frac{4}{3}\pi R^3 \Delta\rho g \tag{1.74}$$

$$v_o = \frac{2}{9}\frac{\Delta\rho g R^2}{\eta_o} \tag{1.75}$$

v_o is the Stokes' velocity and η_o is the viscosity of the medium.
For an O/W emulsion with $\Delta\rho = 0.2$ in water ($\eta_o \sim 10^{-3}$ Pa·s), the rate of creaming or sedimentation is $\sim 4.4 \times 10^{-5}$ m s^{-1} for 10 μm droplets and $\sim 4.4 \times 10^{-7}$ m s^{-1} for 1 μm droplets. This means that in a 0.1 m container creaming or sedimentation of the 10 μm droplets is complete in ~ 0.6 h and for the 1 μm droplets this takes ~ 60 h.

2) Moderately concentrated emulsions ($0.2 < \phi < 0.1$). In this case, one has to take into account the hydrodynamic interaction between the droplets, which

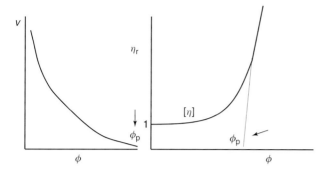

Figure 1.28 Variation of v and η_r with ϕ.

reduces the Stokes velocity to a value v given by the following expression [29]

$$v = v_o(1 - k\varphi) \tag{1.76}$$

where k is a constant that accounts for hydrodynamic interaction. k is of the order of 6.5, which means that the rate of creaming or sedimentation is reduced by about 65%.

3) Concentrated emulsions ($\phi > 0.2$). The rate of creaming or sedimentation becomes a complex function of ϕ as is illustrated in Figure 1.28, which also shows the change of relative viscosity η_r with ϕ.

As can be seen from the above figure, v decreases with increase in ϕ and ultimately it approaches zero when ϕ exceeds a critical value, ϕ_p, which is the so-called "maximum packing fraction." The value of ϕ_p for monodisperse "hard spheres" ranges from 0.64 (for random packing) to 0.74 for hexagonal packing. The value of ϕ_p exceeds 0.74 for polydisperse systems. Also for emulsions that are deformable, ϕ_p can be much larger than 0.74.

The above figure also shows that when ϕ approaches ϕ_p, η_r approaches ∞. In practice, most emulsions are prepared at ϕ values well below ϕ_p, usually in the range 0.2–0.5, and under these conditions, creaming or sedimentation is the rule rather than the exception. Several procedures may be applied to reduce or eliminate creaming or sedimentation, and these are discussed in the following sections.

1.8.2
Prevention of Creaming or Sedimentation

1) **Matching density of oil and aqueous phases**: Clearly, if $\Delta\rho = 0$, $v = 0$; However, this method is seldom practical. Density matching, if possible, only occurs at one temperature.
2) **Reduction of droplet size**: As the gravity force is proportional to R_3, then if R is reduced by a factor of 10, the gravity force is reduced by 1000. Below a certain droplet size (which also depends on the density difference between oil and water), the Brownian diffusion may exceed gravity and creaming or sedimentation is prevented. This is the principle of formulation of nanoemulsions

(with size range 20–200 nm), which may show very little or no creaming or sedimentation. The same applies for microemulsions (size range 5–50 nm)

3) **Use of "thickeners"**: These are high-molecular-weight polymers, natural, or synthetic such as Xanthan gum, hydroxyethyl cellulose, alginates, and carragenans. To understand the role of these "thickeners," let us consider the gravitational stresses exerted during creaming or sedimentation

$$\text{Stress} = \text{mass of drop} \times \text{acceleration of gravity} = \frac{4}{3}\pi R^3 \Delta\rho g \quad (1.77)$$

To overcome such stress, one needs a restoring force

$$\text{Restoring force} = \text{area of drop} \times \text{stress of rop} = 4\pi R^2 \sigma_p \quad (1.78)$$

Thus, the stress exerted by the droplet σ_p is given by

$$\sigma_p = \frac{\Delta\rho R g}{3} \quad (1.79)$$

Simple calculation shows that σ_p is in the range 10^{-3} to 10^{-1} Pa, which implies that for prediction of creaming or sedimentation one needs to measure the viscosity at such low stresses. This can be obtained by using constant stress or creep measurements.

The above-described "thickeners" satisfy the criteria for obtaining very high viscosities at low stresses or shear rates. This can be illustrated from plots of shear stress σ and viscosity η versus shear rate $\dot{\gamma}$ (or shear stress), as shown in Figure 1.29. These systems are described as "pseudoplastic" or shear thinning. The low shear (residual or zero shear rate) viscosity $\eta(0)$ can reach several thousand Pascal-second, and such high values prevent creaming or sedimentation [30, 31].

The above behavior is obtained above a critical polymer concentration (C^*), which can be located from plots of $\log \eta$ versus $\log C$ as shown in Figure 1.30. Below C^*, the $\log \eta - \log C$ curve has a slope in the region of 1, whereas above C^*, the slope of the line exceeds 3. In most cases, good correlation between the rate of creaming or sedimentation and $\eta(0)$ is obtained.

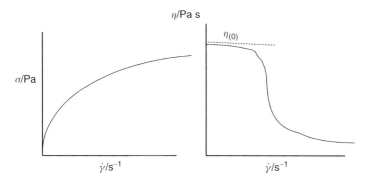

Figure 1.29 Variation of stress σ and viscosity η with shear rate $\dot{\gamma}$.

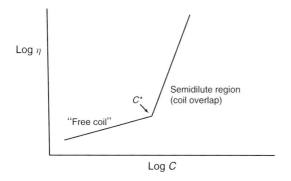

Figure 1.30 Variation of log η with log C for polymer solutions.

4) **Controlled flocculation**: As discussed earlier, the total energy distance of separation curve for electrostatically stabilized shows a shallow minimum (secondary minimum) at relatively long distance of separation between the droplets. By addition of small amounts of electrolyte, such minimum can be made sufficiently deep for weak flocculation to occur. The same applied for sterically stabilized emulsions, which show only one minimum, whose depth can be controlled by reducing the thickness of the adsorbed layer. This can be obtained by reducing the molecular weight of the stabilizer and/or addition of a nonsolvent for the chains (e.g., electrolyte).

The above phenomenon of weak flocculation may be applied to reduce creaming or sedimentation, although in practice this is not easy since one has also to control the droplet size.

5) **Depletion flocculation**: This is obtained by addition of "free" (nonadsorbing) polymer in the continuous phase [32]. At a critical concentration, or volume fraction of free polymer, ϕ_p^+, weak flocculation occurs because the free polymer coils become "squeezed out" from between the droplets. This is illustrated in Figure 1.31 that shows the situation when the polymer volume fraction exceeds the critical concentration.

The osmotic pressure outside the droplets is higher than in between the droplets, and this results in attraction whose magnitude depends on the concentration of the free polymer and its molecular weight, as well as the droplet size and ϕ. The value of ϕ_p^+ decreases with increase in the molecular weight of the free polymer. It also decreases as the volume fraction of the emulsion increases.

The above weak flocculation can be applied to reduce creaming or sedimentation although it suffers from the following drawbacks: (i) temperature dependence; as the temperature increases, the hydrodynamic radius of the free polymer decreases (due to dehydration) and hence more polymer will be required to achieve the same effect at lower temperatures. (ii) If the free polymer concentration is increased above a certain limit, phase separation may occur and the flocculated emulsion droplets may cream or sediment faster than in the absence of the free polymer.

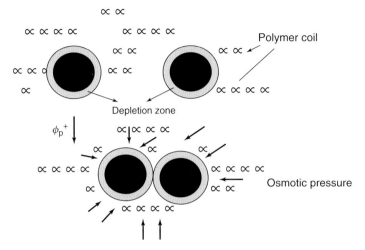

Figure 1.31 Schematic representation of depletion flocculation.

1.9
Flocculation of Emulsions

Flocculation is the result of van der Waals attraction that is universal for all disperse systems. The van der Waals attraction G_A was described before. As shown in Figure 1.6, G_A is inversely proportional to the droplet–droplet distance of separation h and it depends on the effective Hamaker constant A of the emulsion system. One way to overcome the van der Waals attraction is by electrostatic stabilization using ionic surfactants, which results in the formation of electrical double layers that introduce a repulsive energy that overcomes the attractive energy. Emulsions stabilized by electrostatic repulsion become flocculated at intermediate electrolyte concentrations (see below). The second and most effective method of overcoming flocculation is by "steric stabilization" using nonionic surfactants or polymers. Stability may be maintained in electrolyte solutions (as high as 1 mol dm^{-3} depending on the nature of the electrolyte) and up to high temperatures (in excess of 50 °C) provided that the stabilizing chains (e.g., PEO) are still in better than θ-conditions ($\chi < 0.5$).

1.9.1
Mechanism of Emulsion Flocculation

This can occur if the energy barrier is small or absent (for electrostatically stabilized emulsions) or when the stabilizing chains reach poor solvency (for sterically stabilized emulsions, i.e., $\chi > 0.5$). For convenience, flocculation of electrostatically and sterically stabilized emulsions are discussed separately.

1.9.1.1 Flocculation of Electrostatically Stabilized Emulsions

As discussed before, the condition for kinetic stability is $G_{max} > 25\,kT$. When $G_{max} < 5\,kT$, flocculation occurs. Two types of flocculation kinetics may be distinguished: fast flocculation with no energy barrier and slow flocculation when an energy barrier exists.

The fast flocculation kinetics was treated by Smoluchowki [33], who considered the process to be represented by second-order kinetics and the process is simply diffusion controlled. The number of particles n at any time t may be related to the final number (at $t = 0$) n_o by the following expression

$$n = \frac{n_o}{1 + kn_o t} \tag{1.80}$$

where k is the rate constant for fast flocculation that is related to the diffusion coefficient of the particles D, that is

$$k = 8\pi D R \tag{1.81}$$

D is given by the Stokes–Einstein equation

$$D = \frac{kT}{6\pi \eta R} \tag{1.82}$$

Combining Eqs. (1.81) and (1.82)

$$k = \frac{4}{3}\frac{kT}{\eta} = 5.5 \times 10^{-18}\,\text{m}^3\text{s}^{-1} \text{ for water at } 25\,°\text{C} \tag{1.83}$$

The half life $t_{1/2}$ ($n = (1/2)n_o$) can be calculated at various n_o or volume fraction ϕ as given in Table 1.6.

The slow flocculation kinetics was treated by Fuchs [34] who related the rate constant k to the Smoluchowski rate by the stability constant W

$$W = \frac{k_o}{k} \tag{1.84}$$

W is related to G_{max} by the following expression [35]

$$W = \frac{1}{2}\exp\left(\frac{G_{max}}{kT}\right) \tag{1.85}$$

Table 1.6 Half-life of emulsion flocculation.

R (μm)	ϕ			
	10^{-5}	10^{-2}	10^{-1}	5×10^{-1}
0.1	765 s	76 ms	7.6 ms	1.5 ms
1.0	21 h	76 s	7.6 s	1.5 s
10.0	4 mo	21 h	2 h	25 mo

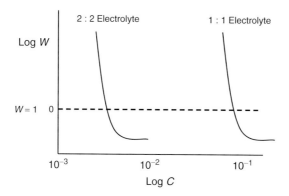

Figure 1.32 Log W–log C curves for electrostatically stabilized emulsions.

As G_{max} is determined by the salt concentration C and valency, one can derive an expression relating W to C and Z

$$\log W = -2.06 \times 10^9 \left(\frac{R\gamma^2}{Z^2}\right) \log C \tag{1.86}$$

where γ is a function that is determined by the surface potential ψ_o

$$\gamma = \left[\frac{\exp(Ze\psi_o/kT) - 1}{\exp(ZE\psi_o/kT) + 1}\right] \tag{1.87}$$

Plots of log W versus log C are shown in Figure 1.32. The condition log $W = 0$ ($W = 1$) is the onset of fast flocculation. The electrolyte concentration at this point defines the critical flocculation concentration (CFC). Above the CFC, $W < 1$ (due to the contribution of van der Waals attraction that accelerates the rate above the Smoluchowski value). Below the CFC, $W > 1$ and it increases with decrease of electrolyte concentration. The figure also shows that the CFC decreases with increase of valency in accordance to the Scultze–Hardy rule.

Another mechanism of flocculation is that involving the secondary minimum (G_{min}), which is few kilotesla units. In this case, flocculation is weak and reversible and hence one must consider both the rate of flocculation (forward rate k_f) and deflocculation (backward rate k_b). The rate of decrease of particle number with time is given by the expression

$$-\frac{dn}{dt} = -k_f n^2 + k_b n \tag{1.88}$$

The backward reaction (breakup of weak flocs) reduces the overall rate of flocculation.

1.9.1.2 Flocculation of Sterically Stabilized Emulsions

This occurs when the solvency of the medium for the chain becomes worse than a θ-solvent ($\chi > 0.5$). Under these conditions, G_{mix} becomes negative, that is, attractive and a deep minimum is produced resulting in catastrophic flocculation (referred to as *incipient flocculation*). This is schematically represented in Figure 1.33.

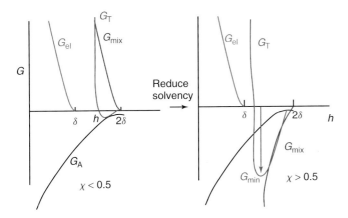

Figure 1.33 Schematic representation of flocculation of sterically stabilized emulsions.

With many systems, good correlation between the flocculation point and the θ-point is obtained.

For example, the emulsion will flocculate at a temperature (referred to as the critical flocculation temperature, CFT) that is equal to the θ-temperature of the stabilizing chain. The emulsion may flocculate at a critical volume fraction of a nonsolvent critical volume fraction (CFV), which is equal to the volume of nonsolvent that brings it to a θ-solvent.

1.9.2
General Rules for Reducing (Eliminating) Flocculation

1) **Charge-stabilized emulsions, for example, using ionic surfactants**: The most important criterion is to make G_{max} as high as possible; this is achieved by three main conditions: high surface or zeta potential, low electrolyte concentration, and low valency of ions.
2) **Sterically stabilized emulsions**: Four main criteria are necessary: (i) complete coverage of the droplets by the stabilizing chains. (ii) Firm attachment (strong anchoring) of the chains to the droplets. This requires the chains to be insoluble in the medium and soluble in the oil. However, this is incompatible with stabilization that requires a chain that is soluble in the medium and strongly solvated by its molecules. These conflicting requirements are solved by the use of A-B, A-B-A block, or BA_n graft copolymers (B is the "anchor" chain and A is the stabilizing chain(s)).

Examples of the B chains for O/W emulsions are polystyrene, polymethylmethacrylate, PPO, and alkyl PPO. For the A chain(s), PEO and polyvinyl alcohol are good examples. For W/O emulsions, PEO can form the B chain, whereas the A chain(s) could be polyhydroxy stearic acid (PHS), which is strongly solvated by most oils. (iii) Thick adsorbed layers: the adsorbed layer thickness should be in the range of 5–10 nm. This means that the molecular weight of the stabilizing chains could be in the range of 1000–5000. (iv) The stabilizing chain should be maintained in good solvent conditions ($\chi < 0.5$) under all conditions of temperature changes on storage.

1.10
Ostwald Ripening

The driving force for Ostwald ripening is the difference in solubility between the small and large droplets (the smaller droplets have higher Laplace pressure and higher solubility than the larger ones). This is illustrated in Figure 1.34 where r_1 decreases and r_2 increases as a result of diffusion of molecules from the smaller to the larger droplets.

The difference in chemical potential between different sized droplets was given by Lord Kelvin [36]

$$S(r) = S(\infty) \exp\left(\frac{2\gamma V_m}{rRT}\right) \tag{1.89}$$

where $S(r)$ is the solubility surrounding a particle of radius r, $S(\infty)$ is the bulk solubility, V_m is the molar volume of the dispersed phase, R is the gas constant, and T is the absolute temperature. The quantity ($2\gamma V_m/RT$) is termed the *characteristic length*. It has an order of ~1 nm or less, indicating that the difference in solubility of a 1 μm droplet is on the order of 0.1% or less. Theoretically, Ostwald ripening should lead to condensation of all droplets into a single drop. This does not occur in practice since the rate of growth decreases with increase of droplet size.

For two droplets with radii r_1 and r_2 ($r_1 < r_2$)

$$\frac{RT}{V_m} \ln\left[\frac{S(r_1)}{S(r_2)}\right] = 2\gamma \left[\frac{1}{r_1} - \frac{1}{r_2}\right] \tag{1.90}$$

Equation (1.88) shows that the larger the difference between r_1 and r_2, the higher the rate of Ostwald ripening.

Ostwald ripening can be quantitatively assessed from plots of the cube of the radius versus time t [37, 38]

$$r^3 = \frac{8}{9}\left[\frac{S(\infty)\gamma V_m D}{\rho RT}\right] t \tag{1.91}$$

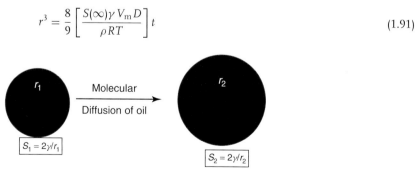

Figure 1.34 Schematic representation of Ostwald ripening.

D is the diffusion coefficient of the disperse phase in the continuous phase and ρ is the density of the disperse phase.

Several methods may be applied to reduce Ostwald ripening [39–41]. (i) Addition of a second disperse phase component that is insoluble in the continuous medium (e.g., squalane). In this case, partitioning between different droplet sizes occurs, with the component having low solubility expected to be concentrated in the smaller droplets. During Ostwald ripening in a two-component system, equilibrium is established when the difference in chemical potential between different size droplets (which results from curvature effects) is balanced by the difference in chemical potential resulting from partitioning of the two components. This effect reduces further growth of droplets. (ii) Modification of the interfacial film at the O/W interface. According to Eq. (1.89), reduction in γ results in reduction of Ostwald ripening rate. By using surfactants that are strongly adsorbed at the O/W interface (i.e., polymeric surfactants) and that do not desorb during ripening (by choosing a molecule that is insoluble in the continuous phase), the rate could be significantly reduced. An increase in the surface dilational modulus $\varepsilon (= d\gamma/d\ln A)$ and decrease in γ would be observed for the shrinking drop, and this tends to reduce further growth.

A-B-A block copolymers such as PHS-PEO-PHS (which is soluble in the oil droplets but insoluble in water) can be used to achieve the above effect. Similar effects can also be obtained using a graft copolymer of hydrophobically modified inulin, namely INUTEC®SP1 (ORAFTI, Belgium). This polymeric surfactant adsorbs with several alkyl chains (which may dissolve in the oil phase) leaving loops and tails of strongly hydrated inulin (polyfructose) chains. The molecule has limited solubility in water and hence it resides at the O/W interface. These polymeric emulsifiers enhance the Gibbs elasticity thus significantly reducing the Ostwald ripening rate.

1.11
Emulsion Coalescence

When two emulsion droplets come in close contact in a floc or creamed layer or during Brownian diffusion, thinning and disruption of the liquid film may occur resulting in eventual rupture. On close approach of the droplets, film thickness fluctuations may occur – alternatively, the liquid surfaces undergo some fluctuations forming surface waves, as illustrated in Figure 1.35.

The surface waves may grow in amplitude and the apices may join as a result of the strong van der Waals attraction (at the apex, the film thickness is the smallest). The same applies if the film thins to a small value (critical thickness for coalescence).

Figure 1.35 Schematic representation of surface fluctuations.

A very useful concept was introduced by Deryaguin and Scherbaker [42] who suggested that a "disjoining pressure" $\pi(h)$ is produced in the film that balances the excess normal pressure

$$\pi(h) = P(h) - P_o \qquad (1.92)$$

where $P(h)$ is the pressure of a film with thickness h and P_o is the pressure of a sufficiently thick film such that the net interaction free energy is zero.

$\pi(h)$ may be equated to the net force (or energy) per unit area acting across the film

$$\pi(h) = -\frac{dG_T}{dh} \qquad (1.93)$$

where G_T is the total interaction energy in the film.

$\pi(h)$ is made of three contributions due to electrostatic repulsion (π_E), steric repulsion (π_s), and van der Waals attraction (π_A)

$$\pi(h) = \pi_E + \pi_s + \pi_A \qquad (1.94)$$

To produce a stable film $\pi_E + \pi_s > \pi_A$, this is the driving force for prevention of coalescence that can be achieved by two mechanisms and their combination: (i) increased repulsion both electrostatic and steric and (ii) dampening of the fluctuation by enhancing the Gibbs elasticity. In general, smaller droplets are less susceptible to surface fluctuations and hence coalescence is reduced. This explains the high stability of nanoemulsions.

Several methods may be applied to achieve the above effects:

1) **Use of mixed surfactant films**: In many cases using mixed surfactants, for example, anionic and nonionic or long-chain alcohols, can reduce coalescence as a result of several effects such as igh Gibbs elasticity, high surface viscosity, and hindered diffusion of surfactant molecules from the film.
2) **Formation of lamellar liquid crystalline phases at the O/W interface**: This mechanism was suggested by Friberg and coworkers [43], who suggested that surfactant or mixed surfactant film can produce several bilayers that "wrap" the droplets. As a result of these multilayer structures, the potential drop is shifted to longer distances thus reducing the van der Waals attraction. A schematic representation of the role of liquid crystals is shown in Figure 1.36 that illustrates the difference between having a monomolecular layer and a multilayer as is the case with liquid crystals.

For coalescence to occur, these multilayers have to be removed "two-by-two" and this forms an energy barrier preventing coalescence.

1.11.1
Rate of Coalescence

As film drainage and rupture is a kinetic process, coalescence is also a kinetic process. If one measures the number of particles n (flocculated or not) at time t

$$n = n_t + n_v m \qquad (1.95)$$

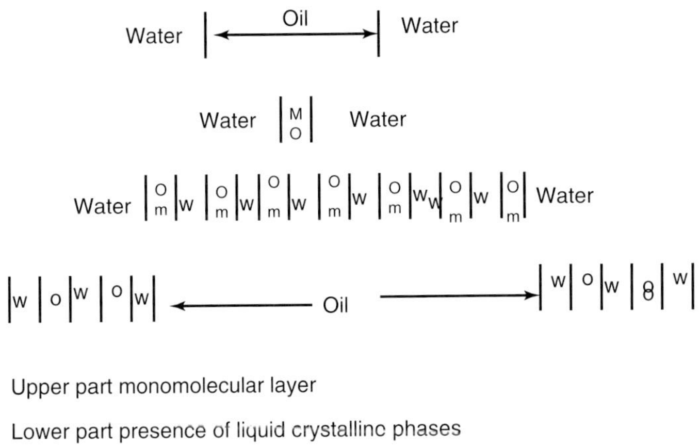

Water |← Oil →| Water

Water |M/O| Water

Water |O/m|W|O/m|W|O/m|W|O/m|W|O/m|W W/m|O|W|O/m| Water

|W|O|W|O|W| ← Oil → |W|O|W|8|W|

Upper part monomolecular layer
Lower part presence of liquid crystalline phases

Figure 1.36 Schematic representation of the role of liquid crystalline phases.

where n_t is the number of primary particles remaining and n is the number of aggregates consisting of m separate particles.

For studding emulsion coalescence, one should consider the rate constant of flocculation and coalescence. If coalescence is the dominant factor, then the rate K follows a first-order kinetics

$$n = \frac{n_o}{Kt}\left[1 + \exp(-Kt)\right] \quad (1.96)$$

which shows that a plot of log n versus t should give a straight line from which K can be calculated.

1.11.2
Phase Inversion

Phase inversion of emulsions can be one of two types: transitional inversion induced by changing the facers that affect the HLB of the system, for example, temperature and/or electrolyte concentration and catastrophic inversion, which is induced by increasing the volume fraction of the disperse phase.

Catastrophic inversion is illustrated in Figure 1.37 that shows the variation of viscosity and conductivity with the oil volume fraction ϕ. As can be seen, inversion occurs at a critical ϕ, which may be identified with the maximum packing fraction. At ϕ_{cr}, η suddenly decreases – the inverted W/O emulsion has a much lower volume fraction. κ also decreases sharply at the inversion point because the continuous phase is now oil.

Earlier theories of phase inversion were based on packing parameters. When ϕ exceeds the maximum packing (~0.64 for random packing and ~0.74 for hexagonal packing of monodisperse spheres; for polydisperse systems, the maximum packing exceeds 0.74), inversion occurs. However, these theories are not adequate, because many emulsions invert at ϕ values well below the maximum packing as a result of the change in surfactant characteristics with variation of conditions. For example,

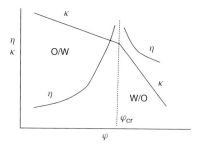

Figure 1.37 Variation of conductivity and viscosity with volume fraction of oil.

when using a nonionic surfactant based on PEO, the latter chain changes its solvation by increase of temperature and/or addition of electrolyte. Many emulsions show phase inversion at a critical temperature (the PIT) that depends on the HLB number of the surfactant as well as the presence of electrolytes. By increasing temperature and/or addition of electrolyte, the PEO chains become dehydrated and finally they become more soluble in the oil phase. Under these conditions, the O/W emulsion will invert to a W/O emulsion. The above dehydration effect amounts to a decrease in the HLB number and when the latter reaches a value that is more suitable for W/O emulsion inversion will occur. At present, there is no quantitative theory that accounts for phase inversion of emulsions.

1.12
Rheology of Emulsions

The rheology of emulsions has many similar features to that of suspensions. However, they differ in three main aspects. (i) The mobile liquid/liquid interface that contains surfactant or polymer layers introduces a response to deformation and one has to consider the interfacial rheology. (ii) The dispersed-phase viscosity relative to that of the medium has an effect on the rheology of the emulsion. (iii) The deformable nature of the dispersed-phase droplets, particularly for large droplets, has an effect on the emulsion rheology at high phase volume fraction ϕ.

When the above factors are considered, one can treat the bulk rheology of emulsions in a similar way as for suspensions and the same techniques used can be applied.

1.12.1
Interfacial Rheology

1.12.1.1 Interfacial Tension and Surface Pressure

A fluid interface in equilibrium exhibits an intrinsic state of tension that is characterized by its interfacial tension γ, which is given by the change in free energy with area of the interface, at constant composition n_i and temperature T

$$\gamma = \left(\frac{\partial G}{\partial A}\right)_{n_i,T} \quad (1.97)$$

The unit for γ is energy per unit area ($\mathrm{mJm^{-2}}$) or force per unit length ($\mathrm{mNm^{-1}}$), which are dimensionally equivalent.

Adsorption of surfactants or polymers lowers the interfacial tension and this produces a two-dimensional surface pressure π that is given by

$$\pi = \gamma_o - \gamma \quad (1.98)$$

where γ_o is the interfacial tension of the "clean" interface (before adsorption) and γ that after adsorption.

1.12.1.2 Interfacial Shear Viscosity

The interface is considered to be a macroscopically planer, dynamic fluid interface. Thus, the interface is regarded as a two-dimensional entity independent of the surrounding three-dimensional fluid. The interface is considered to correspond to a highly viscous insoluble monolayer and the interfacial stress σ_s acting within such a monolayer is sufficiently large compared to the bulk fluid stress acting across the interface, and in this way, one can define an interfacial shear viscosity η_s

$$\sigma_s = \eta_s \dot{\gamma} \quad (1.99)$$

where $\dot{\gamma}$ is the shear rate. η_s is given in surface pascal seconds ($\mathrm{N\,m^{-1}\,s}$) or surface poise (dyne per centimeter second).

It should be mentioned that the surface viscosity of a surfactant-free interface is negligible and it can reach high values for adsorbed rigid molecules such as proteins.

1.12.2
Measurement of Interfacial Viscosity

Many surface viscometers use torsional stress measurements on rotating a ring, disk, or knife-edge (Figure 1.38) within or near to the liquid/liquid interface [44]. This type of viscometer is moderately sensitive; for a disk viscometer, the interfacial shear viscosity can be measured in the range $\eta_s \geq 10^{-2}$ surface Pa s.

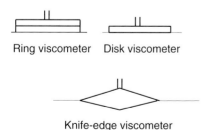

Figure 1.38 Schematic representation of surface viscometers.

The disk is rotated within the plane of the interface with angular velocity ω. A torque is exerted on the disk of radius R by both the surfactant film with surface viscosity η_s and the viscous liquid (with bulk viscosity η) that is given by the expression

$$M = (8/3)R^3\eta\omega + 4\pi R^2\eta_s\omega \qquad (1.100)$$

1.12.3
Interfacial Dilational Elasticity

The interfacial dilational (Gibbs) elasticity ε that is an important parameter in determining emulsion stability (reduction of coalescence during formation) is given by the following equation:

$$\varepsilon = \frac{d\gamma}{d\ln A} \qquad (1.101)$$

where $d\gamma$ is the change in interfacial tension during expansion of the interface by an amount dA (referred to as *interfacial tension gradient* resulting from nonuniform surfactant adsorption on expansion of the interface).

One of the most convenient methods for measurement of ε is to use a Langmuir trough with two moving barriers for expansion and compression of the interface. Another method for measurement of ε is to use the oscillating bubble technique and instruments are commercially available.

A useful method for measurement of ε is the pulsed drop method [45]. Rapid expansion of a droplet at the end of the capillary from a radius r_1 to r_2 is obtained by application of pressure. The pressure drop within the droplet is measured as a function of time using a sensitive pressure transducer. From the pressure drop, one can obtain the interfacial tension as a function of time. The Gibbs dilational elasticity is determined from values of the time-dependent interfacial tension. Measurement can be made as a function of frequency as illustrated in Figure 1.39 for stearic acid at the decane–water interface at pH $= 2.5$.

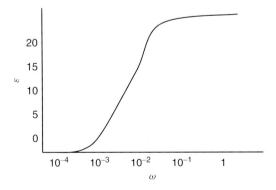

Figure 1.39 Gibbs dilational elasticity versus frequency.

1.12.4
Interfacial Dilational Viscosity

Measurement of the dilational viscosity is more difficult than measurement of the interfacial shear viscosity. This is due to the coupling between dilational viscous and elastic components. The most convenient method for measurement of dilational viscosity is the maximum bubble pressure technique that can be only applied at the A/W interface. According to this technique, the pressure drop across the bubble surface at the instant when the bubble possesses a hemispherical shape (corresponding to the maximum pressure) is due to a combination of bulk viscous, surface tension, and surface dilational viscosity effects, and this allows one to obtain the interfacial dilational viscosity.

1.12.5
Non-Newtonian Effects

Most adsorbed surfactant and polymer coils at the O/W interface show non-Newtonian rheological behavior. The surface shear viscosity η_s depends on the applied shear rate, showing shear thinning at high shear rates. Some films also show Bingham plastic behavior with a measurable yield stress.

Many adsorbed polymers and proteins show viscoelastic behavior and one can measure viscous and elastic components using sinusoidally oscillating surface dilation. For example, the complex dilational modulus ε^* obtained can be split into an "in-phase" (the elastic component ε') and "out-of-phase" (the viscous component ε'') components. Creep and stress relaxation methods can be applied to study viscoelasticity.

1.12.6
Correlation of Emulsion Stability with Interfacial Rheology

1.12.6.1 Mixed Surfactant Films
Prins et al. [46] found that a mixture of SDS and dodecyl alcohol give a more stable O/W emulsion when compared to emulsions prepared using SDS alone. This enhanced stability is due to the higher interfacial dilational elasticity ε for the mixture when compared to that of SDS alone. Interfacial dilational viscosity did not play a major role because the emulsions are stable at high temperature whereby the interfacial viscosity becomes lower.

The above correlation is not general for all surfactant films because other factors such as thinning of the film between emulsion droplets (which depends on other factors such as repulsive forces) can also play a major role.

1.12.6.2 Protein Films
Biswas and Haydon [47] found some correlation between the viscoelastic properties of protein (albumin or arabinic acid) films at the O/W interface and the stability of emulsion drops against coalescence. Viscoelastic measurements were carried

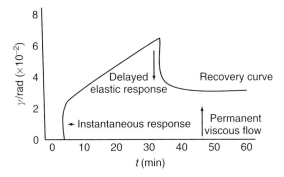

Figure 1.40 Creep curve for protein film at the O/W interface.

out using creep and stress relaxation measurements (using a specially designed interfacial rheometer). A constant torque or stress σ (mN m^{-1}) was applied, and the deformation γ was measured as a function of time for 30 min. After this period, the torque was removed and γ (which changes sign) was measured as a function of time to obtain the recovery curve. The results are illustrated in Figure 1.40.

From the creep curves, one can obtain the instantaneous modulus $G_o(\sigma/\gamma_{int})$ and the surface viscosity η_s from the slope of the straight line (which gives the shear rate) and the applied stress. G_o and η_s are plotted versus pH as shown in Figure 1.41. Both show increase with increase in pH reaching a maximum at $\sim$pH = 6 (the isoelectric point of the protein) at which the protein molecules show maximum rigidity at the interface.

The stability of the emulsion was assessed by measuring the residence time t of several oil droplets at a planer O/W interface containing the adsorbed protein. Figure 1.41 shows the variation of $t_{1/2}$ (time taken for half the number of oil droplets to coalesce with the oil at the O/W interface) with pH. Good correlation between $t_{1/2}$ and G_o and η_s is obtained.

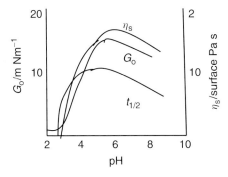

Figure 1.41 Variation of $t_{1/2}$ and G_o and η_s with pH.

Biswas and Haydon [47] derived a relationship between coalescence time τ and surface viscosity η_s, instantaneous modulus G_o, and adsorbed film thickness h

$$\tau = \eta_s \left[3C' \frac{h^2}{A} - \frac{1}{G_o} - \phi(t) \right] \quad (1.102)$$

where $3C'$ is a critical deformation factor, A is the Hamaker constant, and $\phi(t)$ is the elastic deformation per unit stress.

Equation (1.102)) shows that τ increases with increase of η_s, but most importantly, it is directly proportional to h^2. These results show that viscoelasticity is necessary but not sufficient to ensure stability against coalescence. To ensure stability of an emulsion, one must make sure that h is large enough and film drainage is prevented.

1.13
Bulk Rheology of Emulsions

For rigid (highly viscous) oil droplets dispersed in a medium of low viscosity such as water, the relative viscosity η_r of a dilute (volume fraction $\phi \leq 0.01$) O/W emulsion of noninteracting droplets behaves as "hard spheres" (similar to suspensions) [48, 49].

In the above case, η_r is given by the Einstein equation [50]

$$\eta_r = 1 + [\eta]\phi \quad (1.103)$$

where $[\eta]$ is the intrinsic viscosity that is equal to 2.5 for hard spheres.

For droplets with low viscosity (comparable to that of the medium), the transmission of tangential stress across the O/W interface, from the continuous phase to the dispersed phase, causes liquid circulation in the droplets. Energy dissipation is less than that for hard spheres and the relative viscosity is lower than that predicted by the Einstein equation.

For an emulsion with viscosity η_i for the disperse phase and η_o for the continuous phase [48]

$$[\eta] = 2.5 \left(\frac{\eta_i + 0.4\eta_o}{\eta_i + \eta_o} \right) \quad (1.104)$$

Clearly when $\eta_i \gg \eta_o$, the droplets behave as rigid spheres and $[\eta]$ approaches the Einstein limit of 2.5. In contrast if $\eta_i \ll \eta_o$ (as is the case for foams), $[\eta] = 1$.

In the presence of viscous interfacial layers, Eq. (1.104) is modified to take into account the surface shear viscosity η_s and surface dilational viscosity μ_s

$$[\eta] = 2.5 \left(\frac{\eta_i + 0.4\eta_o + \xi}{\eta_i + \eta_o + \xi} \right) \quad (1.105)$$

$$\xi = \frac{(2\eta_s + 3\mu_s)}{R} \quad (1.106)$$

R is the droplet radius.

When the volume fraction of droplets exceeds the Einstein limit, that is, $\phi > 0.01$, one must take into account the effect of Brownian motion and interparticle interactions. The smaller the emulsion droplets, the more important the contribution of Brownian motion and colloidal interactions. Brownian diffusion tends to randomize the position of colloidal particles, leading to the formation of temporary doublets, triplets, and so on. The hydrodynamic interactions are of longer range than the colloidal interactions, and they come into play at relatively low volume fractions ($\phi > 0.01$) resulting in ordering of the particles into layers and tending to destroy the temporary aggregates caused by the Brownian diffusion. This explains the shear thinning behavior of emulsions at high shear rates.

For the volume fraction range $0.01 < \phi < 0.2$, Bachelor [51] derived the following expression for a dispersion of hydrodynamically interacting hard spheres

$$\eta_r = 1 + 2.5\phi + 6.2\phi^2 + \vartheta\phi^3 \qquad (1.107)$$

The second term in Eq. (1.107) is the Einstein limit, the third term accounts for hydrodynamic (two-body) interaction, while the fourth term relates to multibody interaction.

At higher volume fractions ($\phi > 0.2$), η_r is a complex function of ϕ and the $\eta_r - \phi$ curve is schematically shown in Figure 1.42. This curve is characterized by two asymptotes $[\eta]$ the intrinsic viscosity and ϕ_p the maximum packing fraction.

A good semiempirical equation that fits the curve is given by Dougherty and Krieger [52, 53]

$$\eta_r = \left(1 - \frac{\phi}{\phi_p}\right)^{-[\eta]\phi_p} \qquad (1.108)$$

1.13.1
Analysis of the Rheological Behavior of Concentrated Emulsions

When considering the rheology of concentrated emulsions (without deformation of the emulsion drops), one should attempt to find an expression for the fourth term in ϕ^3 of Eq. (1.107). Unfortunately, there is no theoretical rigorous treatment of this term and only semiempirical equations are available [54, 55] for the case of intermediate volume fractions.

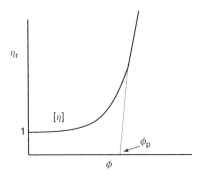

Figure 1.42 $\eta_r - \phi$ curve.

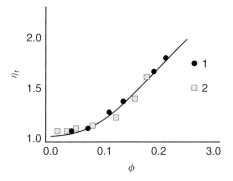

Figure 1.43 Comparison of experimental data of the concentration dependence of different emulsions with some theoretical predictions.

Two models were proposed by Pal [56] that are described by the following expressions

$$\eta_r \left[\frac{2\eta_r + 5\lambda}{2 + 5\lambda} \right]^{1/2} = \exp\left[\frac{2.5\varphi}{1 - (\varphi/\varphi^*)} \right] \quad (1.109)$$

$$\eta_r \left[\frac{2\eta_r + 5\lambda}{2 + 5\lambda} \right]^{1/2} = \left[1 - (\varphi/\varphi^*) \right]^{-2.5\varphi^*} \quad (1.110)$$

where λ is the ratio of viscosities of disperse drops and continuous medium and ϕ^* is the limit of closest packing of drops in free space (as in suspensions), although it was used as a free fitting factor.

The above models describe rather well experimental data for various real emulsions in a wide concentration range as illustrated in Figure 1.43.

The increase of the concentration of drops in emulsions results not only in an increase in viscosity at low shear rates (the limiting residual Newtonian viscosity $\eta(0)$) but also in the appearance of strong non-Newtonian effects, that is, a shear rate dependence of the apparent viscosity. This is illustrated in Figure 1.44 that shows the variation of viscosity with applied stress [54]. This figure shows the remarkable transition from an almost Newtonian behavior at low stresses to an anomalous flow with pronounced non-Newtonian effects.

Another example of the changes in the character of rheological properties just close to the upper boundary of the concentration domain, that is, when approaching the state of closest packing of spherical drops is shown in Figure 1.45 for a water-in-oil emulsion [57, 58].

As seen in Figures 1.44 and 1.45, the approach to the limit of high concentration and transition beyond the closest packing of nondeformable spherical drops leads to principle changes in the rheological properties. The Newtonian viscous flow is replaced by a viscoplastic behavior with jumplike decrease (by several orders of magnitude) in the apparent viscosity in a narrow range of applied stresses. The jump in the apparent viscosity at some shear stress is the reflection of the rupture of the structure, and this stress may be treated as a yield stress.

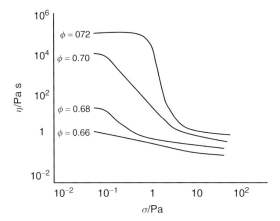

Figure 1.44 Flow curves of model "oil-in-water" emulsion (average drop size of 4.6 μm) at various volume fractions.

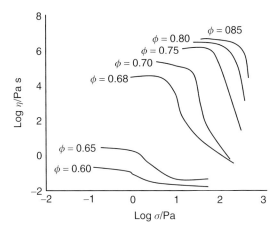

Figure 1.45 Flow curves of "water-in-oil" emulsions when approaching the concentration limit corresponding to the closest packing of spherical drops.

At high volume fractions ($\phi > 0.6$), there is a significant effect of the average droplet diameter (volume to surface ratio d_{32}). The drop size influences the volume to surface ratio, and this leads to a more pronounced effect of the flow inside the drops. This phenomenon becomes more significant when approaching to the upper boundary of intermediate concentrations. This is illustrated in Figures 1.46 and 1.47 that show the variation of viscosity with volume fraction for emulsions with various droplet diameters [56–58]. In the low-volume fraction regime (Figure 1.46), that is, at $\phi < 0.6$, there is hardly any effect of the droplet diameter on the viscosity of the emulsion. However, in the high-volume fraction regime ($\phi > 0.6$), reduction in droplet diameter results in a significant increase of the viscosity.

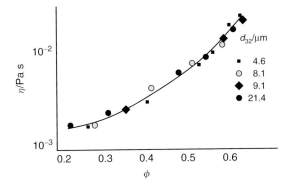

Figure 1.46 Viscosity volume fraction curves for emulsions with different droplet diameters and at low-volume fractions ($\phi < 0.6$).

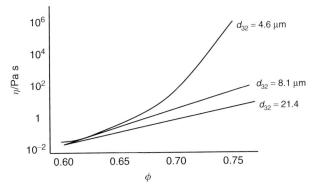

Figure 1.47 Viscosity–volume fraction curves for emulsions with different droplet and at high-volume fractions ($\phi < 0.6$).

Another rheological behavior of concentrated emulsions is the presence of thixotropy. The interfacial layers in the closely arranged drops can produce a certain type of structure, which is destroyed by deformation but restored at rest.

Another rheological behavior of concentrated emulsions is the presence of thixotropy. The interfacial layers in the closely arranged drops can produce some kind of structure, which is destroyed by deformation and restores at rest. The interaction between drops and evolution of their shape in flow can also result in viscoelastic effects as is discussed below.

1.14
Experimental $\eta_r - \phi$ Curves

Experimental results of $\eta_r - \phi$ curves were obtained for paraffin O/W emulsions stabilized with an A-B-C surfactant consisting of nonyl phenol (B), 13 mol propylene oxide stabilized with the surfactant containing 27 EO (the volume medium diameter of the droplets is 3.5 µm). The calculations based on the Dougherty–Krieger equation are also shown in the same figure.

1.14.1
Experimental $\eta_r - \phi$ Curves

Experimental results for $\eta_r - \phi$ curves were obtained for paraffin O/W emulsions [48] stabilized with an A-B-C surfactant consisting of nonyl phenol (B), 13 mol of propylene oxide (C), and PEO with 27, 48, 80, and 174 mol EO. As an illustration, Figure 1.48 shows the results for an emulsion stabilized with the surfactant containing 27 EO 9 (the volume median diameter of the droplets was 3.5 µm). Calculations based on the Dougherty–Krieger equation are also shown in the same figure. In these calculations, $[\eta] = 2.5$ and ϕ_p was obtained from a plot of $\eta^{-1/2}$ versus ϕ and extrapolation of the straight line to $\eta^{-1/2} = 0$. The value of ϕ_p was 0.73 (which is higher than the maximum random packing of 0.64) as a result of the polydispersity of the emulsion. The results using the other three surfactants showed the same trend; the experimental $\eta_r - \phi$ curves are close to those calculated using the Dougherty–Krieger equation indicating that these emulsions behave as hard spheres.

1.14.2
Influence of Droplet Deformability

The influence of droplet deformability on emulsion rheology was investigated by Saiki *et al.* [59] by comparing the $\eta_r - \phi$ curves of hard spheres of silica with two dimethylsiloxane poly-dimethylsiloxane (PDMS) emulsions with low (PDMS 0.3) and high deformability (PDMS 0.45) (by controlling the proportion of cross-linking agent for the droplets; 0.3 low and 0.45 high cross-linking agent). The $\eta_r - \phi$ curves for the three systems are shown in Figure 1.49. The $\eta_r - \phi$ curve for silica can be

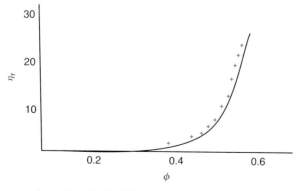

+ experimental results; Full line calculated curve using Dougherty-Krieger equation

Figure 1.48 Experimental and theoretical $\eta_r-\phi$ curve.

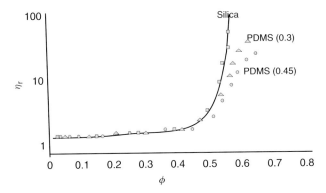

Figure 1.49 η_r–ϕ curves for silica and two PDMS emulsions.

fitted by the Dougherty–Krieger equation over the whole volume fraction range indicating typical hard sphere behavior. The $\eta_r - \phi$ curve for the less deformable PDMS deviates from the hard sphere curve at $\phi = 0.58$. The $\eta_r - \phi$ curve for the more deformable PDMS deviates from the hard sphere curve at $\phi = 0.40$. This clearly shows that the deformation of the "soft" droplets occurs at relatively low-volume fraction.

1.15
Viscoelastic Properties of Concentrated Emulsions

The viscoelastic properties of emulsions can be investigated using dynamic (oscillatory) measurements. A sinusoidal strain with amplitude γ_o is applied on the system at a frequency ω (rad s^{-1}), and the stress σ (with amplitude σ_o) is simultaneously measured. From the time shift Δt between the sine waves of strain and stress, one can measure the phase angle shift δ ($\delta = \Delta t \omega$).

From σ_o, γ_o, and δ, one can obtain the complex modulus G^*, the storage modulus G' (the elastic component), and the loss modulus G'' (the viscous component).

G^*, G', and G'' are measured as a function of strain amplitude to obtain the linear viscoelastic region and then as a function of frequency (keeping γ_o in the linear region). As an illustration, Figure 1.50 shows the results for an O/W emulsion at $\phi = 0.6$ (the emulsion was prepared using an A-B-A block copolymer of PEO, A and PPO, B with an average of 47 PO units and 42 EO units [49].

The results of Figure 1.50 are typical for a viscoelastic liquid. In the low-frequency regime (<1 Hz), $G'' > G'$. As the frequency ω increases, G' increases, at a characteristic frequency ω^* (the cross-over point), G' becomes higher than G'', and at high frequency, it becomes closer to G^*. G'' increases with increase in frequency reaching a maximum at ω^* after which it decreases with further increase in frequency.

From ω^*, one can calculate the relaxation time t^*

$$t* = \frac{1}{2\pi \omega *} \quad (1.111)$$

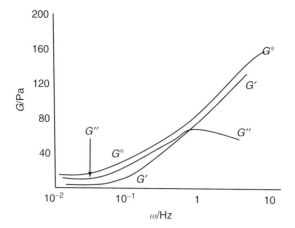

Figure 1.50 Variation of G^*, G', and G'' with frequency ω/Hz.

For the above value of ϕ (= 0.6), $t^* = 0.12$ s. t^* increases with increase in ϕ and this reflects the stronger interaction with increase in ϕ.

To obtain the onset of strong elastic interaction in emulsions, G^*, G', and G'' (obtained in the linear viscoelastic region and high frequency, e.g., 1 Hz) are plotted versus the volume fraction of the emulsion ϕ. One should make sure that the droplet size distribution in all emulsions is the same. The most convenient way is to prepare an emulsion at the highest possible ϕ (e.g., 0.6), and this emulsion is then diluted to obtain various ϕ values. Droplet size analysis should be obtained for each emulsion to make sure that the size distribution is the same.

Figure 1.51 shows the plots for G^*, G', and G'' versus ϕ. At $\phi < 0.56$, $G'' > G'$ whereas at $\phi > 0.56$, $G' > G'' - \phi = 0.56$ is the onset of predominantly elastic interaction and this reflects the small distance of separation between the droplets.

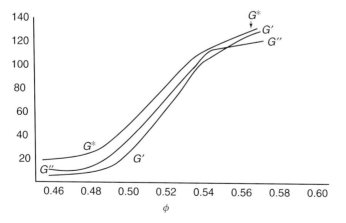

Figure 1.51 Variation of G^*, G', and G'' with ϕ.

1.15.1
High Internal Phase Emulsions (HIPEs)

The maximum packing fraction ϕ^* of nondeformable droplets in an emulsion is in the region of 0.71–0.75, depending on the droplet size distribution and arrangement of drops in space. However, some emulsion systems can exceed this maximum, that is, with $\phi > \phi^*$ and these are referred to as high internal phase emulsions (HIPEs). To achieve this, deformation of the spherical droplets must take place via compression of a dispersion resulting in transformation of the spherical droplets into tightly packed polygon-shaped particles that occupy the space. These systems have wide application in cosmetics, food stuffs, and emulsion explosives.

The general thermodynamic approach to understanding the nature and properties of HIPEs was proposed by Princen [60]. According to this approach, HIPEs are created by application of outer pressure that compresses the drops and transform them from spheres to polygons. This outer pressure is equivalent to the osmotic pressure Π acting inside the thermodynamic system. The work produced by this pressure when creating an HIPE is equal to the stored energy by the increase of droplet surface area S due to changes in shape. This equality is given by the following expression:

$$-\Pi dV = \sigma dS \qquad (1.112)$$

where σ is the interfacial tension.

Equation (1.112)) shows that the osmotic pressure for decreasing the volume ΠdV is equal to the work needed for creating additional new surface dS.

Substituting the expression for concentration gives the equation for the osmotic pressure as a function of volume fraction ϕ and change in surface area S (reduced by the volume V)

$$\Pi = \sigma \varphi^2 \frac{d(S/V)}{d\varphi} \qquad (1.113)$$

The stored surface energy serves as a source of elasticity of the HIPE, which is observed in shear deformation [61–64]. The experimental evidence of this conception is seen in close correlation between the concentration dependence of the shear elastic modulus G and osmotic pressure Π as shown in Figure 1.52. The experimental data of Figure 1.52 are reduced by the Laplace pressure (σ/R).

Using a reduction factor, (σ/R) reflects the proposed conception of elasticity of HIPEs as a consequence of the increase of surface energy on compression of a drop [61–63]. This approach presumes that both G and Π are inversely proportional to droplet size. The concentration dependence of elasticity should be the product $\phi^{1/3}(\phi - \phi^*)$ or $\phi(\phi - \phi^*)$ as discussed by Princen and Kiss [65]. The solidlike properties of HIPES can be observed when $\phi > \phi^*$. The elasticity of HIPEs can be illustrated from measurement of the modulus as a function of frequency. This is shown in Figure 1.53 for a model emulsion of monodisperse droplets ($R = 500$ nm) of poly(dimethyl siloxane) in water at a volume fraction ϕ of 0.98 [65].

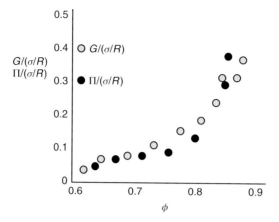

Figure 1.52 Correlation between the elastic modulus (open circles) and osmotic pressure (closed circles) for HIPEs.

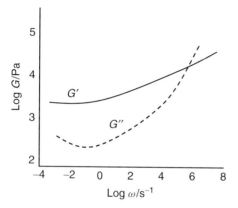

Figure 1.53 Variation of the storage modulus G' and loss modulus G'' with frequency for poly(dimethyl siloxane) emulsion ($r = 600$ nm) at $\phi = 0.98$.

The elastic modulus G' shows little dependency on frequency up to 10^2 s^{-1} indicating that the HIPE behaves as a linear elastic material. However, G' increases at very high frequencies and this effect is attributed to a mechanical glass transition of the emulsion as a viscoelastic material [65].

Further rheological measurements [66] indicated that HIPEs show nonlinear viscoelastic as well as viscous behavior. This is illustrated in Figure 1.54 that shows the variation of the complex modulus with stress for highly concentrated emulsions at various volume fractions. It can be seen that the modulus remains virtually constant with the increase in stress, but at a critical stress, it shows a rapid decrease indicating "softening" of the structure at high stress values. Such behavior is typical for "structured" colloidal dispersions that undergo destruction of this structure when the stress exceeds a critical value.

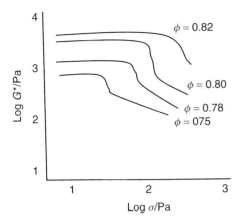

Figure 1.54 Variation of complex modulus with applied stress for highly concentrated (cosmetic grade) water-in-oil emulsions.

It is important to compare the amplitude dependence of the elastic (storage) modulus G' and the viscous (loss) modulus G'' under conditions of large deformation. As mentioned earlier, HIPEs behave as elastic systems with $G' > G''$. However, at high amplitudes, a solidlike to liquidlike behavior is observed and at a critical strain γ^* (to be referred to as the *melting strain*) $G' = G''$ and this can be considered as a measure of the point of rupture of the materials structure. Above γ^*, $G'' > G'$.

Some authors [67] observed structure formation with increasing strain amplitude, a phenomenon analogous to negative (anti-) thixotropy.

The non-Newtonian behavior of HIPEs can also be demonstrated from flow curves [68] as illustrated in Figure 1.55 for W/O emulsion (liquid emulsion explosive) at various volume fractions of HIPEs. As can be seen in Figure 1.55, the yield values show a large increase with a small increase in the volume fraction of the emulsions.

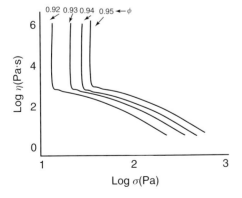

Figure 1.55 Flow curves of highly concentrated water-in-oil emulsions (liquid explosives).

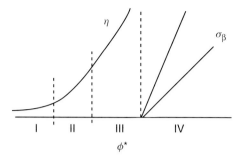

Figure 1.56 General trends of rheological properties of emulsions in the whole volume fraction range.

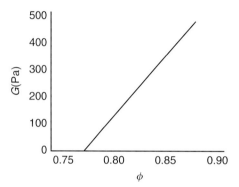

Figure 1.57 Variation of elastic modulus G with volume fraction ϕ.

A schematic representation of the evolution of rheological properties of emulsions from dilute ($\phi \ll 1$) to highly concentrated emulsions ($\phi > \phi^*$) is shown in Figure 1.56 [58]. The transition into the domain of highly concentrated emulsions is accompanied by change in the volume fraction dependence of the rheological properties and the influence of droplet size. This is illustrated in Figures 1.57 and 1.58, which show the variation of elastic modulus G and yield value σ_β with volume fraction ϕ at values above ϕ^*. This linear dependence of G and σ_β on ϕ is consistent with Princen and Kiss [65] discussed earlier.

The influence of droplet size on the viscosity of concentrated emulsions was investigated by Pal [69] who showed that the viscosity of smaller droplets is higher than that of larger droplets at the same volume fraction. This is illustrated in Figure 1.59, which shows the flow curves for an emulsion with $\phi = 0.76$ at two droplet sizes of 12 and 30 μm. In addition to the higher viscosity of the emulsion with the smaller size, the latter shows a more pronounced non-Newtonian effect when compared with the emulsion with the larger size.

The dependence of elastic modulus on droplet diameter can be approximated by the following equation:

$$G = a d_{32}^{-2} \tag{1.114}$$

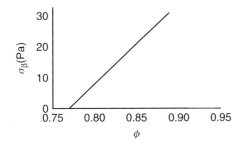

Figure 1.58 Variation of yield stress σ_β with volume fraction ϕ.

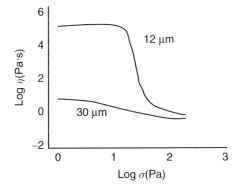

Figure 1.59 Flow curves for concentrated emulsion ($\phi = 0.75$) with two different droplet sizes.

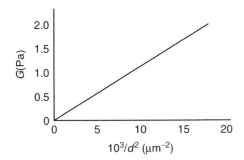

Figure 1.60 Dependence of elastic modulus on average droplet size for highly concentrated emulsions.

Equation (1.114) shows that a plot of G versus $(1/d^2)$ should give a straight line as illustrated in Figure 1.60.

It is worth mentioning that according to the generally accepted Princen [60], Mason et al. [63] theory of the dependence of G on d was always considered as reciprocal linear (but not squared) as it follows from the basic concept of elasticity of HIPEs discussed previously.

1.15.2
Deformation and Breakup of Droplets in Emulsions during Flow

During flow, the emulsion drops undergo deformation (from spherical to ellipsoidal shape), which is then followed by break up to smaller drops. The driving force for drop deformation is the shear stress, and such deformation is resisted by the interfacial tension as determined by the Laplace pressure. Thus, the morphology of a drop is determined by the ratio of stress to the Laplace pressure, that is, the capillary number Ca given by the following expression:

$$\text{Ca} = \frac{\eta_o \dot{\gamma}}{\sigma/R} \tag{1.115}$$

where η_o is the viscosity of the medium, $\dot{\gamma}$ is the shear rate, σ is the interfacial tension, and R is the droplet radius.

The degree of anisotropy D of a deformed drop is based on the classical Taylor model for the viscosity of dilute emulsions [70]

$$D = \frac{16 + 19\lambda}{16(\lambda + 1)} \text{Ca} \tag{1.116}$$

where λ is the ratio of viscosity of disperse phase and disperse medium.

For a moderately concentrated emulsion, one must take into account the dynamic interaction between drops and D is given by the following expression [71]:

$$D = \left[\frac{16 + 19\lambda}{16(\lambda + 1)}\right]\left[1 + \frac{5(2 + 5\lambda)}{4(\lambda + 1)}\varphi\right]\text{Ca} \tag{1.117}$$

A successful method for obtaining experimental results at various emulsion volume fractions is based on modification of the capillary number whereby the viscosity of the medium η_o is replaced by the "mean field" viscosity, that is, the viscosity of the emulsion as a whole η_{em}

$$\text{Ca}_m = \frac{\eta_{em} \dot{\gamma}}{\sigma/R} \tag{1.118}$$

A plot of D versus Ca_m is shown in Figure 1.61 for emulsions with different volume fractions. All results fall on the same line confirming the validity of Eq. (1.118).

The connection between the shape of the droplet and the whole complex behavior of emulsions was established in a series of publications [71–77] for various flow geometries. The final results were obtained in an analytical form. The shear stress in steady flow is expressed as a function of shear rate and capillary number.

The shear stress τ is related to Ca, ratio of viscosities of the disperse phase and medium λ, and volume fraction ϕ of the oil by the following expression:

$$\tau = \frac{2K\text{Ca}f_1 f_2^2}{3(\text{Ca}^2 + f_1^2)} \tag{1.119}$$

where f_1 and f_2 are given by the following expressions:

$$f_1 = \frac{40(\lambda + 1)}{(3 + 2\lambda)(16 + 9\lambda)} \tag{1.120}$$

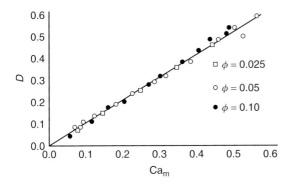

Figure 1.61 Variation of D with Ca$_m$ for emulsions with various volume fractions.

$$f_2 = \frac{5}{3 + 2\lambda} \qquad (1.121)$$

The factor K represents the influence of volume fraction on the viscosity

$$K = \left(\frac{6\sigma}{5R}\right) \frac{(\lambda + 1)(3 + \lambda)\varphi}{5(\lambda + 1) - 5(2 + 5\lambda)\varphi} \qquad (1.122)$$

The problem of calculating the droplet deformation in a flow of viscous liquid was rigorously formulated by Maffettone and Minale [78]. This deformation consists of the transition from spherical to ellipsoidal shape. The exact solution of this problem was given by Wetzel and Tucker [79] (without taking into account the interfacial tension) and later by Jackson and Tuker [80] who proposed a complete solution including the influence of all factors affecting the shape of a drop. A comparison between the theoretical prediction of the dependence of D on Ca and the experimental results [80] is shown in Figure 1.62.

The deformation of drops in flow from spherical to ellipsoidal shape influences the viscosity of the emulsion [81]. This is confirmed by measurement of the viscosity of an emulsion (water in viscous alkyd resin) at various shear rates. In the low

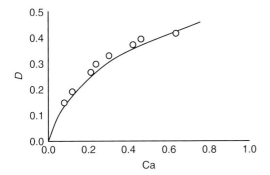

Figure 1.62 Comparison of theoretical prediction of dependence of droplet deformation on capillary number in viscous liquid (solid line) with experimental results (circles).

shear rate regime (where no deformation occurs), the viscosity–volume fraction curve is very close to that of a suspension. However, in the high shear rate regime where deformation of the drops occurs as a result of the low viscosity ratio λ, a non-Newtonian flow is observed and the volume fraction dependence of viscosity is given by the following empirical equation:

$$\eta = \eta_0(1 - \varphi) \tag{1.123}$$

Equation (1.123) shows that the viscosity of the emulsion in the high shear rate regime is lower than that of the medium.

The above problem of calculating the deformation of a drop in a flow is considered without taking into account inertia, that is, at very low Reynolds number. Estimations show that the increase of Reynolds number enhances the impact of inertia, which in turn leads to stronger deformation of the drop and consequently to the growth of stresses in the interfacial layer [82]. It also influences the stability of the drop, which is determined by surface stresses.

The possibility of drop breakup is determined by the balance of the outer stress created by the flow of liquid around the drop (given by the product of the viscosity and shear rate $\eta_0 \dot{\gamma}$) and the Laplace pressure (σ/R). Thus, the determining factor for drop stability is a critical value for the capillary number Ca* that depends on the ratio of the viscosities of disperse droplets and medium λ. The value of Ca* decreases with increase of λ in the domain $\lambda < 1$ and Arcivos [83] expressed the variation of Ca* with λ (at low values) by

$$\text{Ca}^* = 0.054 \lambda^{-2/3} \tag{1.124}$$

Complete results were obtained by Grace [84] who examined both simple shear and two-dimensional extension in the full range of λ values as illustrated in Figure 1.63.

Figure 1.63 shows two interesting results, namely, a minimum in Ca* of 0.4 when $\lambda = 1$ (i.e., when the viscosities of the disperse phase and medium are equal) and the absence of drop break-up in laminar flow when $\lambda > 4$ (i.e., drops of high viscosity).

The results of systematic investigation of single droplet breakup are shown in Figure 1.64. The experiential data are in reasonable agreement with the theoretical

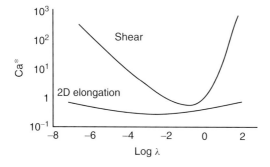

Figure 1.63 Dependence of Ca* on λ in simple shear and two-dimensional extensional flow.

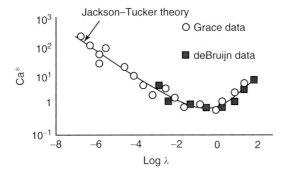

Figure 1.64 Correlation of theoretical dependence of Ca* on λ (solid line) with experimental data (circles and squares) in laminar simple shear.

prediction of Jackson and Tucker [80]. The *breakup conditions* are defined as the limit of their deformation as discussed above. It is assumed that when a deformation results in some steady state of a drop, then this rate of deformation is less than that corresponding to the critical value Ca*. The calculations show that droplet deformation becomes continuous without limit, which means that the drop breaks when Ca > Ca*.

The critical conditions for droplet breakup in a viscoelastic medium are different from a purely viscous liquid. Surface stresses at the interface can vary and are a function of the Reynolds number and the Weissenberg number (ratio of characteristic time of outer action and inner relaxation). Numerical modeling demonstrated that the capillary number increases with increasing the Weissenberg number, that is, enhancement of the viscoelasticity of the medium [85].

The above discussion refers to the case of single drops. However, in concentrated emulsions (practical systems), modifications are needed to take into account the droplet–droplet interactions. A convenient method is to modify the definition of Ca* and λ by substituting the viscosity of the medium by that of the emulsion (mean field approximation). In this way, the modified viscosity ratio λ_m is given by

$$\lambda_m = \frac{\eta_{dr}}{\eta_{em}} \quad (1.125)$$

The results of experimental studies discussed in terms of the function $Ca_m^*(\lambda_m)$ are shown in Figure 1.65 for emulsions with a wide volume fraction range (up to $\phi = 0.7$). The influence of volume fraction is clearly shown in Figure 1.66, which shows plots of the critical shear rate of breakup as a function of reciprocal radius. The higher the value of ϕ, the lower the shear rate required for breakup of the drops. This is consistent with the increase of stress with increasing ϕ.

The drop breakup at a given shear rate can continue up to a limiting value R_{lim} because the capillary number decreases with the decrease of radius and finally it becomes less than the critical value Ca*. This is illustrated in Figure 1.67, which shows the dependence of R_{lim} on shear rate. A parabolic relationship is obtained

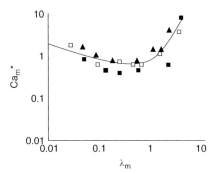

Figure 1.65 Condition of breakup for silicone oil-in-water emulsions at different oil volume fractions (0–0.7).

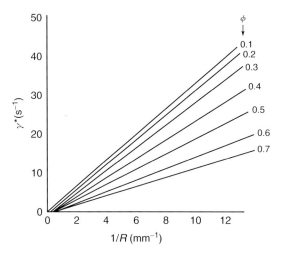

Figure 1.66 Dependence of the critical shear rate corresponding to break up on droplet size for emulsions at different oil volume fractions.

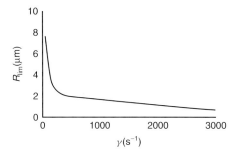

Figure 1.67 Dependence of droplet size (as a result of laminar shearing) on the shear rate for silicone oil-in-water emulsions ($\phi = 0.7$).

that is represented by the following scaling law [86]:

$$R_{\lim} = C \frac{\sigma}{\eta \dot{\gamma}} \qquad (1.126)$$

The factor C is of the order 1 and this reflects the critical value of the capillary number.

It is essential that all theoretical models and experiential results must be under conditions of laminar flow. The transition to higher Reynolds number (>2000), that is, under turbulent regime makes the picture of breakup of droplets in emulsions more complicated. The basic problem is the presence of large fluctuations of local velocities and stresses, which makes the theoretical analysis much more complicated when compared to the case of laminar flow.

Generally speaking, there are two regimes for turbulent flow, namely, turbulent inertial (TI) and "turbulent viscous" (TV). The differences between them are related to the ratio of characteristic sizes of liquid droplets and the turbulent vortex [87]. The minimum droplet size in the TI regime depends on the ratio of the dynamic fluctuation (breakup of a droplet) and surface tension, while for TV regime, the breakup of droplets occur under shear stresses across the continuous medium.

Vankova et al. [88] showed that the maximum size of a droplet in the TI regime, $d_{TI,max}$ is given by the following expression:

$$d_{TI,max} = A_1 \left(\varepsilon^{-2/5} \sigma^{3/5} \rho_c^{-3/5} \right) = A_1 d_k \qquad (1.127)$$

where A_1 is a factor that is of the order 1, ε is the intensity of energy dissipation characterizing the dynamic situation in a flow, and ρ_c is the density of the continuous phase. The term in brackets designated as d_k is a characteristic length.

The maximum size of a drop in the TV regime, $d_{TV,max}$, is determined by the viscous shear stresses

$$d_{TV,max} = A_2 (\varepsilon^{-1} \eta_o^{-1/2} \rho_c^{-1/2} \sigma) \qquad (1.128)$$

where the constant $A_2 \approx 4$ and η_o is the viscosity of the medium.

Equation (1.128)) is only valid for low viscosity drops. For emulsions with more viscous drops dispersed in a medium of arbitrary viscosity, $d_{TV,max}$ is given by the following general expression [89–93]:

$$d_{TX,max} = A_3 \left(1 + A_4 \frac{\eta_{dr} \varepsilon^{1/3} d_{TV,max}^{1/3}}{\sigma} \right)^{3/5} d_k \qquad (1.129)$$

where A_3 and A_4 are constants and η_{dr} is the viscosity of the dispersed liquid drops.

The results of experimental investigations of the dependence of droplet size on the determining factors for the TI regime confirm the validity of Eq. (1.30) with $A_1 = 0.86$. This is shown in Figure 1.68 for hexadecane-in-water emulsions using different emulsifiers, where $d_{TI,max}$ is plotted versus d_k. Comparison of experimental results with theory for TV regime is shown in Figure 1.69 for a large number of emulsions again confirming the validity of Eq. (1.129).

The above analysis is focused on the final equilibrium state of the droplets. However, the kinetics of the breakup process is also of great interest. This kinetic

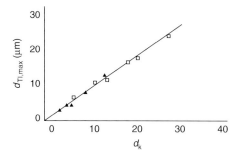

Figure 1.68 Dependence of the maximum droplet diameter $d_{TI,max}$ (micrometer) in turbulent inertial regime on the determining factor d_k predicted by Eq. (1.32) for emulsions with different emulsifiers.

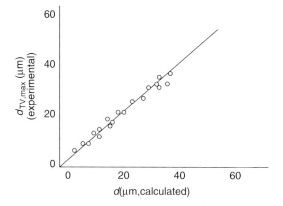

Figure 1.69 Comparison of experimental and theoretical dependence of the maximum droplet diameter $d_{TV,max}$ (micrometer) in turbulent regime on the calculated d predicted by Eq. (1.129) for different emulsions.

process was considered by Vankova et al. [94] who introduced a single additional constant k_{dr} that depends on droplet diameter d

$$k_{dr}(d) = B_1 \frac{\varepsilon^{1/3}}{d^{2/3}} \exp\left[-B_2 \left(\frac{d_k}{d}\right)^{5/3} \left(1 + B_3 \frac{\eta_{dr}\varepsilon^{1/3}d^{1/3}}{\sigma}\right)\right] \quad (1.130)$$

where B_1, B_2, and B_3 are fitting constants.

The experiments carried out on different emulsions confirmed the validity of Eq. (1.33) and allowed Vankova et al. [94] to calculate the values of the constants in Eq. (1.33).

It should be mentioned that the breakup of drops in the flow of emulsions leads to the formation of a large number of droplets with different sizes. The emulsion should be characterized by its maximum size as well as its size distribution. In most cases, the size distribution is represented by a Gaussian function but the real droplet size distribution depends on the viscosity of the droplets [95].

References

1. Tadros, Th.F. and Vincent, B. (1983) in *Encyclopedia of Emulsion Technology* (ed. P. Becher), Marcel Dekker, New York.
2. Binks, B.P. (ed.) (1998) *Modern Aspects of Emulsion Science*, The Royal Society of Chemistry Publication.
3. Tadros, T. (2005) *Applied Surfactants*, Wiley-VCH Verlag GmbH, Germany.
4. Hamaker, H.C. (1937) *Physica (Utrecht)*, **4**, 1058.
5. Deryaguin, B.V. and Landua, L. (1941) *Acta Physicochem. USSR*, **14**, 633.
6. Verwey, E.J.W. and Overbeek, J.Th.G. (1948) *Theory of Stability of Lyophobic Colloids*, Elsevier, Amsterdam.
7. Napper, D.H. (1983) *Polymeric Stabilisation of Dispersions*, Academic Press, London.
8. Walstra, P. and Smolders, P.E.A. (1998) in *Modern Aspects of Emulsions* (ed. B.P. Binks), The Royal Society of Chemistry, Cambridge.
9. Stone, H.A. (1994) *Ann. Rev. Fluid Mech.*, **226**, 95.
10. Wierenga, J.A., ven Dieren, F., Janssen, J.J.M., and Agterof, W.G.M. (1996) *Trans. Inst. Chem. Eng.*, **74-A**, 554.
11. Levich, V.G. (1962) *Physicochemical Hydrodynamics*, Prentic-Hall, Englewood Cliffs, NJ.
12. Davis, J.T. (1972) *Turbulent Phenomena*, Academic Press, London.
13. Lucasses-Reynders, E.H. (1996) in *Encyclopedia of Emulsion Technology* (ed. P. Becher), Marcel Dekker, New York.
14. Graham, D.E. and Phillips, M.C. (1979) *J. Colloid Interface Sci.*, **70**, 415.
15. Lucasses-Reynders, E.H. (1994) *Colloids Surf.*, **A91**, 79.
16. Lucassen, J. (1981) in *Anionic Surfactants* (ed. E.H. Lucassesn-Reynders), Marcel Dekker, New York.
17. van den Tempel, M. (1960) *Proc. Int. Congr. Surf. Act.*, **2**, 573.
18. Griffin, W.C. (1949) *J. Cosmet. Chem.*, **1**, 311; (1954) **5**, 249.
19. Davies, J.T. (1959) *Proc. Int. Congr. Surf. Act.*, **1**, 426.
20. Davies, J.T. and Rideal, E.K. (1961) *Interfacial Phenomena*, Academic Press, New York.
21. Shinoda, K. (1967) *J. Colloid Interface Sci.*, **25**, 396.
22. Shinoda, K. and Saito, H. (1969) *J. Colloid Interface Sci.*, **30**, 258.
23. Beerbower, A. and Hill, M.W. (1972) *Am. Cosmet. Perfum.*, **87**, 85.
24. Hildebrand, J.H. (1936) *Solubility of Non-electrolytes*, 2nd edn, Reinhold, New York.
25. Hansen, C.M. (1967) *J. Paint Technol.*, **39**, 505.
26. Barton, A.F.M. (1983) *Handbook of Solubility Parameters and Other Cohesive Parameters*, CRC Press, New York.
27. Israelachvili, J.N., Mitchell, J.N., and Ninham, B.W. (1976) *J. Chem. Soc., Faraday Trans. 2*, **72**, 1525.
28. Tadros, Th.F. (1967) in *Solid/Liquid Dispersions* (ed. Th.F. Tadros), Academic Press, London.
29. Batchelor, G.K. (1972) *J. Fluid Mech.*, **52**, 245.
30. Buscall, R., Goodwin, J.W., Ottewill, R.H., and Tadros, Th.F. (1982) *J. Colloid Interface Sci.*, **85**, 78.
31. Tadros, T. (2010) *Rheology of Dispersions*, Wiley-VCH Verlag GmbH, Germany.
32. (a) Asakura, S. and Osawa, F. (1954) *J. Phys. Chem.*, **22**, 1255; (b) Asakura, S. and Osawa, F. (1958) *J. Polym. Sci.*, **33**, 183.
33. Smoluchowski, M.V. (1927) *Z. Phys. Chem.*, **92**, 129.
34. Fuchs, N. (1936) *Z. Phys.*, **89**, 736.
35. Reerink, H. and Overbeek, J.Th.G. (1954) *Discuss. Faraday Soc.*, **18**, 74.
36. Thompson, W. (Lord Kelvin) (1871) *Philos. Mag.*, **42**, 448.
37. Lifshitz, I.M. and Slesov, V.V. (1959) *Sov. Phys. JETP*, **35**, 331.
38. Wagner, C. (1961) *Z. Electrochem.*, **35**, 581.
39. Kabalanov, A.S. and Shchukin, E.D. (1992) *Adv. Colloid Interface Sci.*, **38**, 69.
40. Kabalanov, A.S. (1994) *Langmuir*, **10**, 680.
41. Weers, J.G. (1998) in *Modern Aspects of Emulsion Science* (ed. B.P. Binks), Royal Society of Chemistry Publication, Cambridge.
42. Deryaguin, B.V. and Scherbaker, R.L. (1961) *Kolloidn. Zh.*, **23**, 33.

43. Friberg, S., Jansson, P.O., and Cederberg, E. (1976) *J. Colloid Interface Sci.*, **55**, 614.
44. Criddle, D.W. (1960) The viscosity and viscoelasticity of interfaces, in *Rheology*, vol. **3**, Chapter 11 (ed. F.R. Eirich), Academic Press, New York.
45. Edwards, D.A., Brenner, H., and Wasan, D.T. (1991) *Interfacial Transport Processes and Rheology*, Butterworth-Heinemann, Boston, MA, London.
46. Prince, A., Arcuri, C., and van den Tempel, M. (1967) *J. Colloid Interface Sci.*, **24**, 811.
47. Biswas, B. and Haydon, D.A. (1963) *Proc. Roy. Soc.*, **A271**, 296; (1963) **A2**, 317; Biswas, B. and Haydon, D.A. (1962) *Kolloidn. Zh.*, **185**, 31; (1962) **186**, 57.
48. Tadros, Th.F. (1991) Rheological properties of emulsion systems, in *Emulsions — A Fundamental and Practical Approach*, NATO ASI Series, Vol. **363** (ed. J. Sjoblom), Kluwer Academic Publishers, London.
49. Tadros, Th.F. (1994) *Colloids Surf.*, **A91**, 215.
50. Einstein, A. (1906) *Ann. Physik.*, **19**, 289; (1911) **34**, 591.
51. Bachelor, G.K. (1977) *J. Fluid Mech.*, **83**, 97.
52. Krieger, I.M. and Dougherty, T.J. (1959) *Trans. Soc. Rheol.*, **3**, 137.
53. Krieger, I.M. (1972) *Adv. Colloid Interface Sci.*, **3**, 111.
54. Pal, R. (2000) *J. Colloid Interface Sci.*, **225**, 359.
55. Phan-Thien, N. and Pharm, D.C. (1997) *J. Non-Newtonian Fluid Mech.*, **72**, 305.
56. Pal, R. (2001) *J. Rheol.*, **45**, 509.
57. Mason, T.G., Bibette, J., and Weitz, D.A. (1996) *J. Colloid Interface Sci.*, **179**, 439.
58. Derkach, S.R. (2009) *Adv. Colloid Interface Sci.*, **151**, 1.
59. Saiki, Y., Horn, R.G., and Prestidge, C.A. (2008) *J. Colloid Interface Sci.*, **320**, 569.
60. Princen, H.M. (1986) *Langmuir*, **2**, 519.
61. Lacasse, M.D., Grest, C.S., Levine, D., Mason, T.G., and Weitz, D.A. (1996) *Phys. Rev. Lett.*, **76**, 3448.
62. Mason, T.G., Lacasse, M.D., Grest, C.S., Levine, D., Bibette, J., and Weitz, D.A. (1997) *Phys. Rev. E*, **56**, 3150.
63. Mason, T.G. (1999) *Curr. Opin. Colloid Interface Sci.*, **4**, 231.
64. Babak, V.C. and Stebe, M.J. (2002) *J. Dispersion Sci. Technol.*, **23**, 1.
65. Princen, H.M. and Kiss, A.D. (1986) *J. Colloid Interface Sci.*, **112**, 427.
66. Ponton, A., Clement, P., and Grossiord, J.L. (2001) *J. Rheol.*, **45**, 521.
67. Zao, G. and Chen, S.B. (2007) *J. Colloid Interface Sci.*, **316**, 858.
68. Masalova, I. (2007) *Colloids J.*, **69**, 185.
69. Pal, R. (1996) *AICHE J.*, **42**, 3181.
70. Taylor, G.I. (1934) *Proc. R. Soc. London, Ser. A*, **146**, 501.
71. Choi, C.J. and Schowalter, W.R. (1975) *Phys. Fluids*, **18**, 420.
72. Palierne, J.F. (1990) *Rheol. Acta*, **29**, 204.
73. Doi, M. and Ohta, T. (1991) *J. Chem. Phys.*, **95**, 1242.
74. Bousmina, M. (1999) *Rheol. Acta*, **38**, 73.
75. Grmela, M., Bousmina, M., and Palierne, J.F. (2000) *Rheol. Acta*, **40**, 560.
76. Bousmina, M., Grmela, M., and Palierne, J.F. (2002) *Rheol. Acta*, **46**, 1381.
77. Bousmina, M., Grmela, M., and Zhou, Ch. (2002) *J. Rheol.*, **46**, 1401.
78. Maffettone, P.L. and Minale, M. (1998) *J. Non-Newtonian Fluid Mech.*, **78**, 227.
79. Wetzel, E.D. and Tucker, C.L. (2001) *J. Fluid Mech.*, **426**, 199.
80. Jackson, N.E. and Tucker, C.L. (2003) *J. Rheol.*, **47**, 659.
81. Torza, S., Cox, R.G., and Mason, S.G. (1972) *J. Colloid Interface Sci.*, **38**, 395.
82. Li, X. and Sarker, K. (2005) *J. Rheol.*, **49**, 1377.
83. Hinch, T.J. and Arcivos, A. (1980) *J. Fluid Mech.*, **98**, 305.
84. Grace, H.P. (1982) *Chem. Eng. Commun.*, **14**, 225.
85. Renardly, Y. (2008) *Rheol. Acta*, **47**, 89.
86. Mason, T.G. and Bibette, J. (1996) *J. Phys. Rev. Lett.*, **77**, 3481.
87. Heinze, J.O. (1955) *AICHE J.*, **1**, 289.
88. Vankova, N., Tcholakova, S., Denkov, N.D., Ivanov, I.B., Vulchev, V.D., and Danner, Th. (2007) *J. Colloid Interface Sci.*, **312**, 363.
89. Podgorska, W. (2006) *Chem. Eng. Sci.*, **61**, 2986.

90. Calabrese, R.V., Chang, T.P.K., and Dang, P.T. (1986) *AICHE J.*, **32**, 657.
91. Wang, C.Y. and Calabrese, R.V. (1986) *AICHE J.*, **32**, 677.
92. Razzaque, M.M., Afacan, A., Lu, Sh., Nandakumar, K., Jacob, H., Masliyah, J.H., and Sanders, R.S. (2003) *Int. J. Multiphase Flow*, **29**, 1451.
93. Eastwood, C.D., Armi, L., and Lasheras, J.C. (2004) *J. Fluid Mech.*, **502**, 309.
94. Vankova, N., Tcholakova, S., Denkov, N.D., and Danner, Th. (2007) *J. Colloid Interface Sci.*, **313**, 612.
95. Tcholakova, S., Vankova, N., Denkov, N.D., and Danner, Th. (2007) *J. Colloid Interface Sci.*, **310**, 570.

2
Emulsion Formation in Membrane and Microfluidic Devices

Goran T. Vladisavljević, Isao Kobayashi, and Mitsutoshi Nakajima

2.1
Introduction

Conventional devices for bulk emulsification are high-pressure valve homogenizers and rotor stator systems (colloid mills, stirred vessels, toothed disk dispersing machines, etc.) [1]. In these devices, a coarse emulsion with large drops is forced through a high shear region near the rotor or through valves or nozzles to induce turbulence and thereby to break up the drops into smaller ones. The disadvantages of this "top-down" approach are that it is not easily possible to control the mean drop size achieved and the range of drop sizes is usually large. This is a consequence of the fluctuating turbulence stresses in these systems and the exposure of the drops to a variable shear field. In addition, encapsulation yields are often very low, because of the high-energy inputs per unit volume applied [1]. Over the past two decades, new microengineering techniques have been developed for the production of emulsion droplets individually (drop-by-drop). These "bottom-up" techniques include ink-jet printing [2], microfluidic routes [3], microchannel emulsification (MCE) [4], and membrane emulsification (ME) [5]. They can afford creation of uniformly sized droplets of a controlled size, shape, and internal morphology.

This chapter reviews latest developments in the formation of emulsions using membrane, microchannel (MC), and microfluidic techniques. The particle polydispersity is expressed in terms of the coefficient of variation (CV) (relative standard deviation) or the relative span factor. For a specific size distribution, the coefficient of variation is defined as $CV = (\sigma/d_{av}) \times 100$, where σ is the standard deviation of the droplet diameters and d_{av} is the mean droplet diameter. The relative span factor is given by $(d_{90} - d_{10})/d_{50}$, where d_{x0} is the diameter corresponding to $x0$ vol% on a relative cumulative droplet diameter distribution curve.

2.2
Membrane Emulsification (ME)

ME is a process that forms emulsions by injecting a pure dispersed phase or premix into the continuous phase through a microporous membrane. In the former case, fine droplets are produced directly at the membrane/continuous phase interface, whereas in the latter case, preexisting large droplets are homogenized by passing through the pores. ME methods and systems are described schematically in Figure 2.1.

2.2.1
Direct Membrane Emulsification

In conventional direct ME (Figure 2.1a), droplets are produced *in situ* by injecting a pure liquid (the dispersed phase) through the membrane into a second immiscible liquid (the continuous phase) [5]. Hydrophobic and hydrophilic membranes are needed to produce water-in-oil (W/O) and oil-in-water (O/W) emulsions, respectively. At low production rates, droplets can be formed in the absence of any shear on the membrane surface, solely by the action of interfacial tension and "push-off" force [8, 9]. However, in order to obtain uniform droplets at higher production rates, shear stress is generated at the membrane surface, usually by recirculation of the continuous phase along the membrane (Figure 2.1a) [5] or by agitation in a stirred

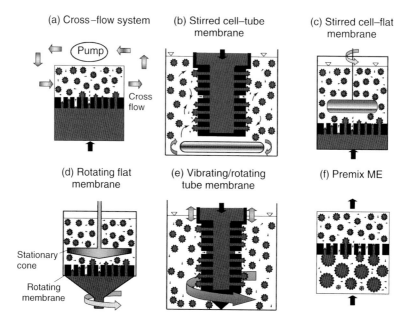

Figure 2.1 Schematic of different membrane emulsification (ME) methods and systems. (Source: Adapted from Refs. [6, 7].)

vessel (Figure 2.1b,c) [10–14]. In a typical tubular ME system, the continuous phase flows inside the membrane, while the pressurized dispersed phase is supplied to the module shell side and permeates across the membrane to the tube side [15]. Cross-flow systems are easy to scale up and offer a constant shear stress along the membrane surface. On the other hand, stirring systems are simpler, easier to operate and a batch volume can be just 10 ml, which can be convenient in some applications, for example, for preparation of emulsions loaded with anticancer drugs [16].

The shear stress can also be generated by a dynamic membrane (Figure 2.1e), in which case the droplet detachment from the membrane surface is facilitated by rotation [17–19] or vibration [20, 21] of the membrane in an otherwise stagnant continuous phase. In a dynamic membrane system, the shear on the membrane surface is decoupled from the cross-flow velocity, meaning that cross flow is applied only to entrain droplets generated in the device and not to provide a surface shear. Thus, very low cross-flow velocities can be used and recirculation of the product stream in a continuous flow system is not needed, which can be advantageous to prevent secondary droplet breakup. Kosvintsev *et al.* [11] modified a commercial Weissenberg "plate and cone" rheometer by replacing normally impervious plate underneath the cone with the membrane and permitting the injection of oil from underneath the membrane into an aqueous phase under constant shear-stress condition (Figure 2.1d). It has been shown that the simple paddle-stirred cell (Figure 2.1d) with nonuniform shear profile on the membrane surface provides the same degree of drop size uniformity as the modified Weissenberg rheometer with a constant shear-stress operation [11]. As a consequence, the modified Weissenberg rheometer was abandoned as a drop formation device and a simple paddle-stirred cell has become increasingly popular as a viable alternative to cross-flow ME systems. Membrane vibration through piezoactuation has been first applied by Zhu and Barrow [20] to provide an extra control over droplet detachment in cross-flow ME. Holdich *et al.* [21] have introduced a vibrating tubular membrane system to control drop generation without any cross flow. Shimoda *et al.* [22] have used a swirl flow to increase throughput in ME by introducing the continuous phase into the membrane tube radially, thereby forming spiral streamlines in the axial direction that exerted a strong centrifugal force on the inner surface of the membrane.

A disadvantage of direct ME is that transmembrane flux should be maintained at relatively low levels ($10-100 \, l \, m^{-2} \, h^{-1}$) to avoid transition from a dripping to continuous outflow regime [23]. Uniformly sized droplets with a CV of around 10% and a span below 0.5 can only be formed in the dripping regime [24–26].

2.2.2
Premix Membrane Emulsification

In premix ME (Figure 2.1f), fine droplets are formed by passing a coarse emulsion through the membrane [27] or porous bed of uniform particles [28]. If the membrane surface is wetted by the continuous phase, a phase inversion may occur during homogenization [29]. Another limiting factor is the pressure difference across the membrane. If the pressure difference is lower than the capillary pressure in a membrane pore, the droplet cannot penetrate the pore, which leads to oil/water separation [30]. In order to achieve additional droplet size reduction and improve droplet size uniformity, emulsion can repeatedly be passed through the same membrane [31–36]. Repeated membrane homogenization was originally developed for homogenization of large multilamellar and unilamellar lipid vesicles using track-etched polycarbonate filters [37]. Premix ME has been reviewed by Nazir *et al.* [38].

2.2.3
Operating Parameters in Membrane Emulsification

The effect of process parameters on droplet generation in ME was reviewed by Joscelyne and Trägårdh [39], Charcosset *et al.* [40], Lambrich and Vladisavljević [41], Gijsbertsen-Abrahamse *et al.* [42], and Yuan *et al.* [43]. The main factors affecting the resultant drop size are wetting properties and microstructure of the membrane (pore size distribution, pore shape, spacing, and tortuosity), but many other parameters play an important role, such as transmembrane flux, shear stress on the membrane surface, viscosity of the continuous and dispersed phase, and surfactant type and concentration. Several models for the prediction of drop size in ME have been developed [44–47].

2.2.4
Membrane Type

Shirasu porous glass (SPG) membrane is the earliest and most common membrane for ME [5, 48]. SPG membrane contains uniform pores with no internal voids or cracks [49], a wide range of commercially available mean pore sizes (0.050–20 µm) [8], and its surface can easily be rendered hydrophobic by chemical modification with organosilane compounds or by coating with silicone resin [50]. SPG membrane is fabricated by phase separation of a primary $Na_2O-CaO-MgO-Al_2O_3-B_2O_3-SiO_2$ type glass, made from a mixture of Shirasu (volcanic ash), calcium carbonate, and boric acid [51]. A hydraulic resistance of conventional SPG membrane is relatively high but can be reduced by making its wall structure asymmetric [52]. Chemical durability of SPG against alkaline solutions can be improved by incorporating ZrO_2 into the structure of SPG membrane [53]. In direct ME, the mean droplet size is three to four times larger than the mean pore size of SPG membrane and a size distribution span under optimal conditions is 0.25–0.45 [54].

In addition to SPG membrane, many other membranes have been investigated for ME including ceramic [54], metallic [55], polymeric [56], and composite.

2.2.4.1 Surfactant Type
The role of surfactant in ME is to rapidly adsorb to the newly formed oil–water interface to reduce the interfacial tension and facilitate the droplet detachment. The effect of dynamic interfacial tension has been investigated by Schröder

et al. [15], Van der Graaf et al. [57], and Rayner et al. [58]. As a rule, the faster the emulsifier molecules adsorb to the newly formed interface, the smaller the droplet size of the resultant emulsion becomes. Surfactant molecules should not adsorb to the membrane surface, because it can cause the dispersed phase to spread over the membrane surface. The effect of emulsifier charge on droplet formation has been investigated by Nakashima et al. [59] for SPG membrane and by Kobayashi et al. [23] for silicon MC plates. It has been found that functional groups of the surfactant molecule must not carry the charge opposite to that of the membrane surface. An untreated SPG membrane has a negative surface potential of -15 to -35 mV within a pH range of 2–8, because of dissociation of acidic silanol groups (Si–OH $\leftrightarrows$ SiO$^-$ + H$^+$). Hence, for this case, the use of cationic emulsifiers such as alkyl-substituted quaternary ammonium salts must be avoided and the same conclusion holds for silicon MCs. Zwitterionic surfactants are also unsuitable, even when they carry a net negative charge. For example, lecithin at pH 3 hinders SPG ME due to electrostatic interactions between positively charged groups (e.g., $-N(CH_3)_3^+$ or $-NH_3^+$) on the phospholipid molecules and negatively charged silanol groups on the SPG membrane, although at pH 3 the net charge of lecithin molecules is negative [60]. Consequently, there is a limit to the type of surfactants that can be used to form droplets using SPG ME. One solution to this limitation is to prepare emulsion using nonionic or anionic surfactant and then to use surfactant-displacement method to subsequently alter the droplet charge [61].

2.2.4.2 Transmembrane Pressure and Wall Shear Stress

The minimum transmembrane pressure for driving the disperse phase through the pores, that is, the capillary pressure name P_{cap}, is given by the Laplace equation

$$P_{cap} = \frac{4\gamma \cos \theta}{d_p} \quad (2.1)$$

where γ is the equilibrium interfacial tension at the disperse phase/continuous phase interface, θ is the contact angle at the interfacial line between the two liquid phases and the membrane surface, and d_p is the mean pore diameter of the membrane. The critical pressure in premix ME is given by Park et al. [32]

$$P_{cap} = \frac{\gamma \left[2 + 2a^6/\sqrt{2a^6 - 1} \times \arccos(1/a^3) - 4a^2 \right]}{a + \sqrt{a^2 - 1}} \quad (2.2)$$

where $a = d_1/d_p$ and d_1 is the mean droplet size in premix. If $d_1/d_p \gg 1$, the capillary pressure is given by Eq. (2.1). In premix ME, the optimum transmembrane pressure is typically 10–50 times greater than the capillary pressure [36]. In direct SPG ME, the operating pressure is usually up to six times above the capillary pressure and the wall shear stress is 2–40 Pa [56]. The higher the wall shear stress, the higher the maximum pressure that can be applied to obtain uniform droplets.

2.3
Microfluidic Junctions and Flow-Focusing Devices

Microfluidics is the science of handling and processing fluids in MCs that have at least one dimension smaller than 1 mm [62]. The first applications of microfluidic technology in blood rheology [63, 64] and chemical analysis [65] were stimulated by the ability of microfluidic devices to use very small amounts of samples and reagents, to carry out analysis in short time, and to achieve high levels of process integration. In the past decade, microfluidic junctions have been widely used for generation of uniform droplets, which is driven by a rising number of applications that can take advantage of precision generation and manipulation of droplets on a microscale [66]. These applications range from capillary electrophoresis [67] and immunoassays [68] to cellomics [69], proteomics [70], and DNA analysis [71]. The most common microfluidic junctions for generation of droplets are T-junction [72–78], ψ-junction [79], cross junction [80], and Y-junction [81]. Microfluidic junctions and flow-focusing devices are usually fabricated by soft lithography [62, 82]. The most commonly used elastomer in soft lithography is poly(dimethylsiloxane) (PDMS), although a number of other polymeric materials can be used, such as polyurethanes [83], polyimides, and phenol formaldehyde polymers.

2.3.1
Microfluidic Junctions

The T-junction is the simplest microfluidic structure for producing and manipulating droplets. The continuous phase is introduced from the horizontal channel, and the dispersed phase flows through the perpendicular channel (Figure 2.2a). The shear forces generated by the continuous phase and the subsequent pressure gradient cause the head of the dispersed phase to elongate into the main channel until the neck of the dispersed phase thins and eventually breaks the stream into a droplet [3, 84]. The dispersed phase should not wet the walls at the junction, for example, hydrophobic T-junctions are required to generate water droplets. Two T-junctions in series with alternating surface wettabilities can be used to produce multiple emulsions with a controlled number of inner drops [85, 86]. When reversing the flow direction, T-junctions with differently sized exit channels will passively sort droplets according to size [87] or break large droplets into uniformly sized smaller droplets, which is known as *geometrically mediated breakup* [73]. In this process, droplets suspended in an immiscible carrier liquid break into two smaller daughter droplets at a T-junction, with each daughter droplet flowing into a separate exit channel (Figure 2.2b). The breakup process can be facilitated by inserting obstructions in exit channels [73]. In the modified T-junction shown in Figure 2.2c, two miscible liquid reagents are supplied from two converging side channels and they mix together before forming a drop in an immiscible continuous phase. Using this strategy, Choi *et al.*, [88] have fabricated calcium alginate beads by injecting alginate and $CaCl_2$ solutions from two different side channels into hexadecane flow in the horizontal channel, where they merge and form droplets.

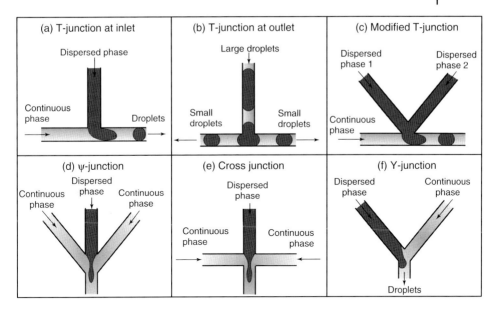

Figure 2.2 Microfluidic junctions for droplet generation and manipulation: (a) T-junction at inlet [72], (b) T-junction at outlet [73, 87], (c) modified T-junction [88], (d) ψ-junction [79], (e) cross junction [80], and (f) Y-junction [81].

Owing to shear forces on the droplet from the continuous phase, the reagents in the droplets are rapidly mixed forming a gel.

At Ψ- or cross junction (Figure 2.2d,e), droplets are generated using a microfluidic extension of Rayleigh's approach [89], with two streams of one liquid flanking a stream of a second immiscible liquid. Droplets are formed when the disperse phase jet becomes too thin to persist in the surrounding continuous phase. Two consecutive Ψ- or cross junctions with alternating wettability can be used to produce droplets or bubbles of core/shell structure [90]. A further extension of this principle to a linear array of three or more cross junctions can allow generation of higher-order multiple emulsions, such as triple, quadruple, and quintuple emulsions [91]. Y-junction shown in Figure 2.2f can be used for drop generation [81], rapid mixing of two miscible liquids, or forming a two-phase flow in a microfluidic channel. The droplet size is independent on the flow rate and viscosity of the dispersed phase [81], which is a behavior different from that in T-junction and flow-focusing device. Y-junction can be used to obtain bichromal particles for "electronic paper" [79, 92] and other anisotropic (Janus) particles.

2.3.2
Microfluidic Flow-Focusing Devices (MFFD)

When droplets are generated in Ψ- or cross junction, the combined two-phase flow is often forced through a small orifice, which is known as *hydrodynamic flow focusing* (Figure 2.3) [93–95]. Figure 2.3a shows a standard microfluidic flow-focusing device

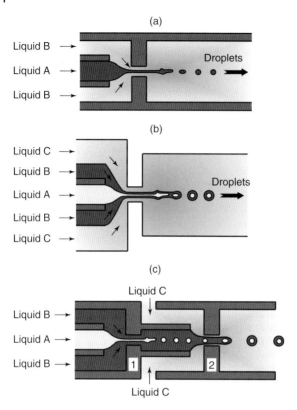

Figure 2.3 Microfluidic flow-focusing devices (MFFD): (a) standard geometry developed by Anna et al. [96] with a constriction placed downstream of three coaxial inlet stream. Liquid B wets channel walls. (b) Modified design with three coaxial streams for generation of core/shell droplets. Liquid C wets channel walls [83]. (c) Two consecutive flow focusing droplet makers. Droplet makers 1 and 2 are wetted by liquid B and C, respectively [97].

(MFFD) developed by Anna et al. [96]: a dispersed phase (liquid A) flows into the middle channel and a continuous phase (liquid B) flows into the two outside channels. The both liquid phases are forced to flow through a small orifice that is located downstream of the three channels. The dispersed phase fluid exerts pressure and shear stress that force the dispersed phase into a narrow thread, which breaks inside or downstream of the orifice. In Figure 2.3a, channel walls are not wetted by liquid A, and thus, hydrophilic and hydrophobic walls are needed to make oil and water droplets, respectively. If liquid A wets the orifice walls, droplets of liquid B will be formed downstream of the orifice. In Figure 2.3b, a coaxial jet composed of two immiscible liquids (A and B) periodically breaks up and forms core/shell droplets in liquid C [83]. In Figure 2.3c, core/shell droplets are formed by employing two consecutive flow-focusing generators with alternating wettability [97]. These droplets can be used as useful templates in the production of hollow polymeric particles [83].

2.4
Microfluidic Devices with Parallel Microchannel Arrays

Flow-focusing devices can generate highly uniform droplets with a CV in the dropping regime of less than 3%. Although the frequency of droplet generation can be as high as 12 000 Hz for W/O droplets [98], the volume flow rate of disperse phase is very low because the droplets are formed from a single MC. In MC array devices, the droplets are formed simultaneously from hundreds or even hundreds of thousands of parallel MCs [99] and the process is called *MCE*. MC arrays can be fabricated onto the surface of MC plate as microgrooves [63] or perpendicular to the plate surface as straight-through MCs [100] (Figure 2.4). First microfluidic device consisting of parallel microgrooves was fabricated by Kikuchi *et al.* [63] using photolithography and anisotropic wet etching in (100) single-crystal silicon (Figure 2.4a).

Photolithography includes masking of substrate with SiO_2 and photoresist, UV exposure, and developing. A channel structure that will be etched into the substrate

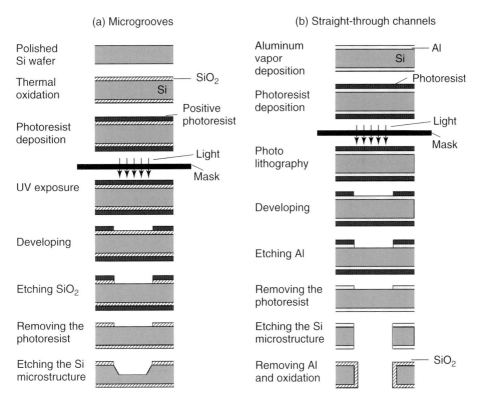

Figure 2.4 Photolithographic methods for fabricating silicon MC array devices: (a) photolithography and anisotropic wet etching for fabrication of shallow microgrooves. (Source: Adapted from Ref. [65]) (b) Photolithography and deep reactive ion etching (DRIE) for fabrication of straight-through channels. (Source: Adapted from Ref. [100].)

is first generated by computer and then drawn onto a transparent plate (mask). A photosensitive polymer (photoresist) is then applied to the substrate in a thin layer, usually by a spin-coating process. If positive photoresist is selected (Figure 2.4a), the patterned areas will be dissolved during the developing step, exposing SiO_2 and the silicon substrate below to the etchant. Wet etching includes (i) etching the SiO_2 layer by hydrofluoric acid at locations unprotected by photoresist; (ii) removing the remaining photoresist, usually by a mixture of sulfuric acid and hydrogen peroxide; and (iii) etching the silicon substrate. Anisotropic etching occurs when etchant (typically a KOH solution) etches silicon at different rates depending on which crystal face is exposed. KOH etches silicon 1–2 orders of magnitude faster than SiO_2, so the SiO_2 layer remains intact. Anisotropic etching is greatly preferred in fabrication of microfluidic devices, because it produces channels with sharp, well-defined edges. Grooved MC arrays can also be fabricated by microcutting in stainless steel [101] and by injection molding in poly(methyl methacrylate) (PMMA) [102]. Stainless steel and PMMA channels are inherently hydrophobic and suitable for generation of W/O emulsions [102].

2.4.1
Grooved-Type Microchannel Arrays

Modules with microgrooves can be either dead end or cross flow. In a typical dead-end module (Figure 2.5a), MCs are fabricated on a terrace and there are four terraces arranged on all four sides of a silicon plate. Each MC is typically 6–12 μm in width, 4–7 μm in depth, and 25–140 μm long [4, 103]. In operation, MC plate

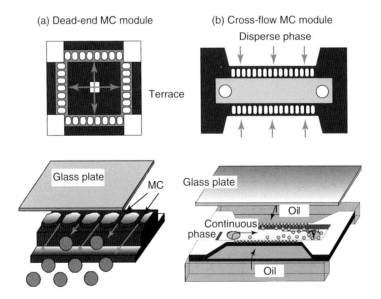

Figure 2.5 Grooved-type MC modules: (a) dead-end module [4]. (b) Cross-flow module [108, 109].

Table 2.1 Typical plate dimensions, number of MCs, average size of resultant droplets and maximum throughput of grooved, and straight-through silicon MC plates.

	Grooved MC cross flow	Grooved MC dead end	Straight-through
Typical size of MC plate	8 × 22.5 mm 60 × 60 mm —	15 × 15 mm 24 × 24 mm 40 × 40 mm	15 × 15 mm 24 × 24 mm 40 × 40 mm
Channel number	Up to 12 000	100–1500	5000–210 000
Average droplet diameter (μm)	1–100	1–100	4–50
Maximum droplet generation frequency per module[a]	$<5 \times 10^5$ droplets s^{-1}	$<8 \times 10^3$ droplets s^{-1}	$<5 \times 10^5$ droplets s^{-1}
Maximum dispersed phase flow rate[a]	1.5 ml h^{-1}	0.1 ml h^{-1}	50 ml h^{-1}

[a] For triglyceride oil-in-water emulsion.

is tightly sealed with a transparent cover plate. The dispersed phase is supplied through a central hole and flows out through MCs on all four sides. MCE exploits the interfacial tension as a driving force for droplet formation. The dispersed phase exiting the MCs takes a disklike shape on the terrace, which is characterized by a higher interfacial area per unit volume than a spherical shape, resulting in hydrodynamic instability. This instability is a driving force for spontaneous transformation of dispersed phase into spherical droplets [104]. Droplet formation behavior drastically changes above the critical velocity, because of transition from dripping to continuous outflow regime. The critical velocity can be predicted from the physical properties of the dispersed and continuous phase, interfacial tension, and system geometry. Different designs of grooved MC arrays have been developed including MCs with partition walls on the terrace [105, 106] and MCs without any terrace [107].

Dead-end modules with grooved MC arrays provide a dispersed phase flow rate of less than 0.1 ml h^{-1} for vegetable oils (Table 2.1), due to limited number of MCs (100–1500). Cross-flow modules with grooved MC arrays are more suited for higher production rates because many parallel cross-flow channels with MC arrays can be incorporated on a single plate [110]. A simplest cross-flow module (Figure 2.5b) has only one cross-flow channel and two holes at its both ends for introduction and withdrawal of the continuous phase. MCs are arranged at both longitudinal sides of the cross-flow channel [108, 109, 111]. The purpose of cross flow is to collect droplets from the module and not to control the droplet size. In the dripping regime, the droplet size is independent on the flow rate of dispersed or continuous phase. In contrast, in flow-focusing devices and T-junctions, the flow rate of all fluid streams has a strong effect on the droplet size. Cross-flow modules with multiple cross-flow channels are available with a maximum size of MC plate of 60 × 60 mm (Table 2.1). This module contains 12 000 microgrooves arranged in 14 parallel arrays and can provide a dispersed phase flow rate of 1.5 ml h^{-1} for soybean O/W emulsions [110].

2.4.2
Straight-through Microchannel Arrays

Grooved-type modules have a limited droplet throughput, because of poor utilization of MC plate surface, because MCs are arranged on the plate surface in longitudinal direction and feed channels for disperse and continuous phase must be provided on the plate surface. A vertical array of straight-through MCs allows much better utilization of the plate surface resulting in significantly higher throughputs (Table 2.1). For example, 60 × 60 mm grooved-type plate with 12 000 MCs can accommodate only 3.3 MCs per 1 mm^2 and provides a maximum soybean oil flow rate of 1.5 ml h^{-1}. On the other hand, 40 × 40 mm straight-through MC plate has 211 248 MCs, that is, 132 MCs per 1 mm^2, and a soybean oil flow rate can exceed 30 ml h^{-1}.

Deep vertical MCs that completely penetrate the substrate are fabricated by deep reactive ion etching (DRIE) [100]. DRIE requires aluminum mask to protect the underlying substrate against etching (Figure 2.4b). DRIE is a three-step process that involves (i) etching a shallow trench into silicon substrate using sulfur hexafluoride (SF_6) plasma, (ii) passivating that newly formed cavity with teflonlike polymer created with the addition of octafluorocyclobutane (C_4F_8) plasma, and (iii) etching a subsequent and deeper trench with SF_6 plasma (Figure 2.5). Passivation with polymer prevents lateral etching of the sidewalls, while the hole becomes deeper. The reactive species (neutral radicals and ions) are formed by the collision of SF_6 molecules with a cloud of energetic electrons excited by an electric field. The reactive species (e.g., F and SF_x^+) react with silicon forming a gaseous substance (SiF_4) that can be removed by a vacuum pump (Figure 2.6).

Straight-through MCs can have either symmetric (Figure 2.7a) or asymmetric (Figure 2.7b) structure. Symmetric MCs are of the same size and shape (e.g., circular or rectangular) along the whole cross section of the plate. Rectangular MCs provide better performance in MCE than circular MCs, and an aspect ratio of slot length to slot width should be at least 3–3.5 to ensure production of

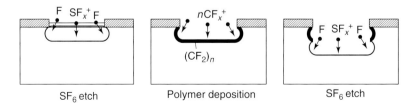

Figure 2.6 Steps in deep reactive ion etching using the Bosch process. The vertically oriented SF_x^+ ions enhance the effect of neutral fluorine radicals in removing the protective film at the bottom of the trench, while the film remains relatively intact along the sidewalls.

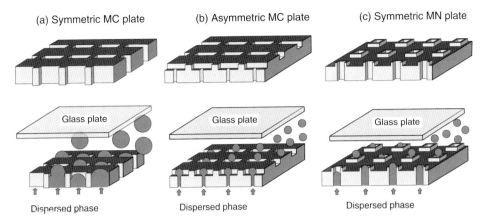

Figure 2.7 Straight-through MC modules: (a) symmetric plate with microslots on both sides [100]. (b) Asymmetric plate with circular channels on upstream side and slots on downstream side [114]. (c) Symmetric plate with micronozzles [121].

uniform droplets [112, 113]. Asymmetric MC plate typically contains cylindrical channels on the upstream (bottom) side and slots on the downstream (top) side (Figure 2.7b). Asymmetric structure is particularly useful for generation of uniform droplets when the dispersed phase viscosity is less than 1 mPa s, for example, when the dispersed phase is a volatile (C6–C10) hydrocarbon, such as decane [114]. Asymmetric straight-through MCs have also been used successfully for production of W/O emulsions [115], polyunsaturated fatty acids (PUFA)-loaded O/W emulsions [116], and *n*-tetradecane emulsions [117, 118]. The size of droplets produced using asymmetric MCs can range from several micrometers [119] to several millimeters [120]. Straight-through micronozzle (MN) array can be employed to increase the velocity of continuous phase at the channel exit, which could be useful if the viscosity of dispersed phase is relatively high [121].

2.5
Glass Capillary Microfluidic Devices

Glass capillary microfluidic devices are pioneered at Harvard by Utada *et al.* [122, 123]. The most critical advantage of these devices is the ability to make truly three-dimensional flow and to achieve versatile multiphase flow configurations. In addition, their wettability can easily be controlled by a surface reaction with an appropriate wetting agent. For example, a treatment with octadecyltrimethoxysilane will make the glass surface hydrophobic, whereas a treatment with 2-[methoxy(polyethyleneoxy)propyl]trimethoxysilane will enhance the hydrophilicity of the glass surface.

Capillary microfluidic devices consist of coaxial assemblies of glass capillaries glued onto a glass slide. First, a circular glass capillary with an outer diameter of about 1 mm is heated and pulled using a micropipette puller. As a result of the pulling process, the capillary breaks into two identical parts, each with a tapered end that culminates in a fine orifice. The orifice is then enlarged to the desired diameter by microforging. The capillary with the desired orifice size is then inserted into a square glass capillary, and the two capillaries are glued together. Coaxial alignment of the two capillaries is ensured by choosing the capillaries such that the outer diameter of the circular capillary is the same as the inner dimensions of the square capillary (Figure 2.8a). The dispersed phase flows inside the circular capillary while the continuous phase flows through the square capillary in the same direction, resulting in a coflow of the two fluids [124]. Under dripping conditions, monodisperse drops are formed at the tip of the capillary [125].

An alternate flow configuration is the countercurrent flow shown in Figure 2.8b. In contrast to coflow device, the two fluids are supplied from the two ends of the same square capillary in opposite directions and both fluids are collected and exit through the circular capillary. The dispersed phase is hydrodynamically flow focused by the continuous phase in the tapered section of the circular capillary,

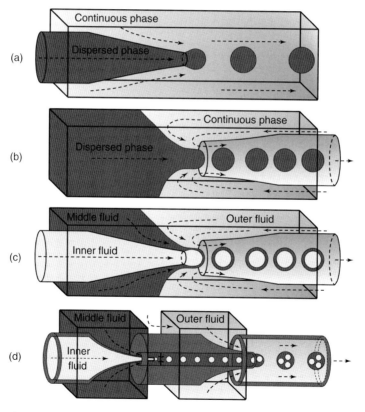

Figure 2.8 Fluid flow configurations in glass capillary devices: (a) coflow of two immiscible fluids [124]. (b) Countercurrent flow of two immiscible fluids [123]. (c) Combination of coflow and countercurrent flow of three immiscible fluids [122]. (d) Two sequential coflow droplet generators [126].

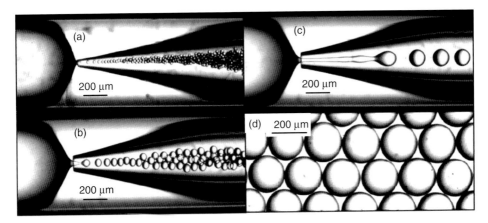

Figure 2.9 Generation of droplets in flow-focusing glass capillary devices: (a) $Q_c = 0.5$ ml h^{-1}, $Q_d = 0.003$ ml h^{-1}, $d_{orifice} = 60$ μm, and $d_{droplet} = 33$ μm. (b) $Q_c = 5$ ml h^{-1}, $Q_d = 1$ ml h^{-1}, $d_{orifice} = 130$ μm, and $d_{droplet} = 100$ μm. (c) $Q_c = 6.5$ ml h^{-1}, $Q_d = 0.7$ ml h^{-1}, $d_{orifice} = 130$ μm, and $d_{droplet} = 230$ μm. Q_d and Q_c are the dispersed and continuous phase flow rate, respectively. (d) Micrograph of collected droplets. The continuous phase is 2 wt% poly(vinyl alcohol) in milli-Q water, and the disperse phase is 5 wt% PLA in DCM [127].

which causes the dispersed phase to break into drops inside the collection tube. Typical micrographs of drop generation process are shown in Figure 2.9a–c. The disperse phase was 5 wt% poly(lactic acid) (PLA) in dichloromethane (DCM), and the continuous phase was 5 wt% poly(vinyl alcohol) in milli-Q water. The drop formation occurs near the orifice in the dripping regime (Figure 2.9a,b) and farther downstream in the jetting regime (Figure 2.9c). The drops formed in the jetting regime are significantly bigger and have a broader size distribution because the point at which a drop separates from the jet can vary. PLA particles were formed after DCM evaporation at room temperature. The production rate in the dripping regime can range from 100 to 7000 Hz [122].

A major advantage of flow focusing is that it can afford to make monodisperse drops with sizes smaller than that of the orifice. This feature is useful because for any given drop size, a capillary with a larger orifice size can be used compared to that in the coflow design, which minimizes the probability of clogging the orifice by the suspended particles or any entrapped debris. Small droplets are formed using capillary with a small orifice under high flow rate of the continuous phase and low flow rate of the disperse phase, that is, for the high values of Q_o/Q_i.

Glass capillary device for making core/shell droplets combines coflow and flow focusing (Figure 2.8c). This device consists of two circular capillaries arranged end to end within a square capillary. The inner fluid is pumped through the tapered circular capillary, while the middle fluid, which is immiscible with the inner and outer fluids, flows through the corners of the outer capillary in the same direction. The outer fluid flows through the square capillary in the opposite direction and hydrodynamically flow focuses the coaxially flowing stream of the other two fluids,

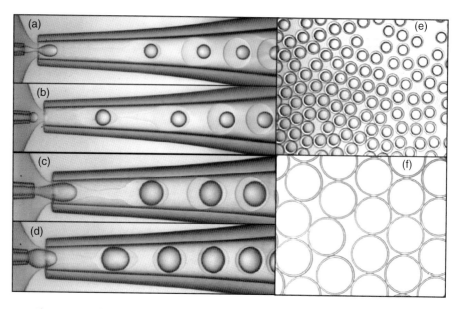

Figure 2.10 The shell thickness of core/shell droplets as a function of the ratio of middle fluid flow rate to inner fluid flow rate: (a) $Q_m/Q_i = 6$. (b) $Q_m/Q_i = 5$. (c) $Q_m/Q_i = 1.5$. (d) $Q_m/Q_i = 0.25$. The diameter of the orifice of the injection and collection tube is 44 and 115 μm, respectively. (e,f) Collected core/shell droplets with a core diameter of 62 μm and a shell thickness of 8 μm. The inner fluid is milli-Q water, the middle fluid is 2 wt% Dow Corning 749 in PDMS (10 cSt), and the outer fluid is 2 wt% poly(vinyl alcohol), 87–89% hydrolyzed, in milli-Q water [132].

which approach from the opposite end. When the three fluids enter the collection tube, double emulsion with core/shell structure is formed under proper flow conditions (Figure 2.10). Typical inner diameters of the tapered end of the injection tube are 10–50 μm, and diameters of the orifice in the collection tube vary from 50 to 400 μm. The drop size and shell thickness can be precisely tuned by adjusting the diameters of the orifices and the flow rates and physical properties of the three fluids. The most important parameter affecting the shell thickness is the ratio of the middle fluid flow rate to the inner fluid flow rate. As shown in Figure 2.10, the higher the Q_m/Q_i ratio, the thicker the shell around the droplet becomes. The ratio of shell thickness to outer drop radius can range from 3 to 40% [122]. The shell material can be polymerized to produce monodisperse polymer shells [128]. Alternatively, a shell may contain dissolved amphiphilic molecules or particles that can undergo self-assembly on solvent evaporation, leading to the generation of vesicles such as liposomes [129], polymersomes [130], and colloidosomes [131].

Device shown in Figure 2.8d employs stepwise emulsification of coflowing streams to create a double emulsion with a constant number of inner drops in each outer drop. The device consists of an injection tube that is inserted into a transition tube. The other end of the transition tube is also tapered and is inserted into a third, coaxially aligned square capillary, the collection tube. The inner fluid flowing through the injection tube is emulsified in the transition tube by coaxial flow of the middle fluid. The single emulsion is subsequently emulsified in the square capillary by coaxial flow of the outer fluid, which is injected into the space between the square capillary and the injection and transition tube. The number of inner drops can be adjusted by controlling the flow rates of the three fluids. Multiple emulsions containing between one and seven inner droplets in each larger droplet have been produced, and the size of both inner and outer drops was precisely controlled [126]. Higher-order multiple emulsions can be made by adding more sequential stages. For example, Chu et al. [126] have made monodisperse triple emulsions by adding a third coflow stage and both the diameter and the number of the individual drops could be efficiently controlled at every level. Chu et al. [126] have demonstrated the ability to form triple emulsions containing between one and seven innermost drops and between one and three middle drops in each outer drop.

2.6
Application of Droplets Formed in Membrane and Microfluidic Devices

The most widespread application of membrane and microfluidic emulsification devices is in the synthesis of monodisperse particles and vesicles with precisely controlled size and internal structure. Vladisavljević and Williams [6] have reviewed the expanding opportunities for the production of particulates using ME. ME is suitable for fabrication of particles with a CV of 10–20%, a controlled size ranging from one micrometer to several hundred micrometers, and from a wide range of materials, including hydrogels [133], polymers [134], inorganic oxides [135], carbon [136], metals [137], and solid lipids [138]. ME can also be used for generation of core/shell droplets if drop generation is followed by internal phase separation [139] or interfacial polymerization [140]. MCE can be used for synthesis of the same particles as ME with an additional benefit that more uniform droplets can be produced (CV < 5%). Droplets produced in microfluidic devices can be used as templates for synthesis of particles with complex geometry and structure, such as nonspherical particles [141] and bifacial (janus) particles [92]. In addition, microfluidic devices permit single-cell encapsulation [142], which enables high-throughput screening of single cells and enzymes for directed evolution.

2.7
Conclusions

Microengineering "bottom-up" strategies for emulsion formation offer a great potential in manufacturing "made-to-measure" droplets with a controlled size and internal structure. Microfluidic junctions and flow-focusing devices can generate single and multiple emulsion droplets in a single step with a polydispersity in the dripping regime of less than 3% and the entrapment efficiency of 100%. MCE can generate emulsions with a CV of less than 5% at the higher production rate than that in microfluidic devices. ME is suitable for fabrication of droplets with a controlled size distribution at even higher productivity, but monodispersity cannot be achieved. Droplets produced in membrane, MC, and microfluidic emulsification devices have been extensively used as templates for fabrication of particles and vesicles. The main benefit of using these devices is in their ability to precisely control the particle size, uniformity, and internal structure.

Acknowledgments

This work was supported by the Engineering and Physical Sciences Research Council (EPSRC) of the United Kingdom (Grant reference number: EP/HO29923/1).

References

1. Karbstein, H. and Schubert, H. (1995) *Chem. Eng. Process.*, **34**, 205–211.
2. Lub, J., Nijssen, W.P.M., Pikkemaat, J.A., and Stapert, H.R. (2006) *Colloid Surf. A*, **289**, 96–104.
3. Teh, S.Y., Lin, R., Hung, L.H., and Lee, A.P. (2008) *Lab Chip*, **8**, 198–220.
4. Kawakatsu, T., Kikuchi, Y., and Nakajima, M. (1997) *J. Am. Oil Chem. Soc.*, **74**, 317–321.
5. Nakashima, T., Shimizu, M., and Kukizaki, M. (1991) *Key Eng. Mater.*, **61-62**, 513–516.
6. Vladisavljević, G.T. and Williams, R.A. (2005) *Adv. Colloid Interface Sci.*, **113**, 1–20.
7. Vladisavljević, G.T. and Williams, R.A. (2008) in *Multiple Emulsions: Technology and Applications* (ed. A. Aserin), John Wiley & Sons, Inc., Hoboken, NJ, pp. 121–164.
8. Kukizaki, M. (2009) *J. Membr. Sci.*, **327**, 234–243.
9. Kosvintsev, S.R., Gasparini, G., and Holdich, R.G. (2008) *J. Membr. Sci.*, **313**, 182–189.
10. Fuchigami, T., Toki, M., and Nakanishi, K. (2000) *J. Sol-Gel Sci. Technol.*, **19**, 337–341.
11. Kosvintsev, S.R., Gasparini, G., Holdich, R.G., Cumming, I.W., and Stillwell, M.T. (2005) *Ind. Eng. Chem. Res.*, **44**, 9323–9330.
12. Ito, F. and Makino, K. (2004) *Colloid Surf. B*, **39**, 17–21.
13. You, J.O., Park, S.B., Park, H.Y., Haam, S., Chung, C.H., and Kim, W.S. (2001) *J. Microencapsul.*, **18**, 521–532.
14. Dragosavac, M.M., Sovilj, M.N., Kosvintsev, S.R., Holdich, R.G., and Vladisavljević, G.T. (2008) *J. Membr. Sci.*, **322**, 178–188.
15. Schröder, V., Behrend, O., and Schubert, H. (1998) *J. Colloid Interface Sci.*, **202**, 334–340.
16. Higashi, S. and Setoguchi, T. (2000) *Adv. Drug Delivery Rev.*, **45**, 57–64.
17. Williams, R.A. (2001) Controlled dispersion using a spinning membrane reactor. UK Patent Application No. PCT/GB00/04917.
18. Vladisavljević, G.T. and Williams, R.A. (2006) *J. Colloid Interface Sci.*, **299**, 396–402.
19. Schadler, V. and Windhab, E.J. (2004) *Chem. Ing. Tech.*, **76**, 1392–1392.
20. Zhu, J. and Barrow, D. (2005) *J. Membr. Sci.*, **261**, 136–144.
21. Holdich, R.G., Dragosavac, M.M., Vladisavljević, G.T., and Kosvintsev, S.R. (2010) *Ind. Eng. Chem. Res.*, **49**, 3810–3817.
22. Shimoda, M., Miyamae, H., Nishiyama, K., Yuasa, T., Noma, S., and Igura, N. (2011) *J. Chem. Eng. J.*, **44**, 1–6.

23. Kobayashi, I., Nakajima, M., and Mukataka, S. (2003) *Colloid Surf. A*, **229**, 33–41.
24. Vladisavljević, G.T., Lambrich, U., Nakajima, M., and Schubert, H. (2004a) *Colloid Surf. A*, **232**, 199–207.
25. Vladisavljević, G.T. and Schubert, H. (2002b) *Desalination*, **144**, 167–172.
26. Vladisavljević, G.T. and Schubert, H. (2003b) *J. Membr. Sci.*, **225**, 15–23.
27. Suzuki, K., Shuto, I., and Hagura, Y. (1996) *Food Sci. Technol. Int. (Tokyo)*, **2**, 43–47.
28. Yasuda, M., Goda, T., Ogino, H., Glomm, W.R., and Takayanagi, H. (2010) *J. Colloid Interface Sci.*, **349**, 392–410.
29. Suzuki, K., Fujiki, I., and Hagura, Y. (1999) *Food Sci. Technol. Int. (Tokyo)*, **5**, 234–238.
30. Koltuniewicz, A.B., Field, R.W., and Arnot, T.C. (1995) *J. Membr. Sci.*, **102**, 193–207.
31. Altenbach-Rehm, J., Schubert, H., and Suzuki, K. (2002) *Chem. Ing. Tech.*, **74**, 587–588.
32. Park, S.H., Yamaguchi, T., and Nakao, S. (2001) *Chem. Eng. Sci.*, **56**, 3539–3548.
33. Yafei, W., Tao, Z., and Gang, H. (2006) *Langmuir*, **67**, 67–73.
34. Vladisavljević, G.T., Shimizu, M., and Nakashima, T. (2006a) *J. Membr. Sci.*, **284**, 373–383.
35. Vladisavljević, G.T., Surh, J., and McClements, D.J. (2006b) *Langmuir*, **22**, 4526–4533.
36. Vladisavljević, G.T., Shimizu, M., and Nakashima, T. (2004b) *J. Membr. Sci.*, **244**, 97–106.
37. Olson, F., Hunt, C.A., and Szoka, F.C. (1979) *Biochim. Biophys. Acta*, **557**, 9–23.
38. Nazir, A., Schroën, K. and Boom, R. (2010) *J. Membr. Sci.*, **362**, 1–11.
39. Joscelyne, S.M. and Trägårdh, G. (2000) *J. Membr. Sci.*, **169**, 107–117.
40. Charcosset, C., Limayem, I., and Fessi, H. (2004) *J. Chem. Technol. Biotechnol.*, **79**, 209–218.
41. Lambrich, U. and Vladisavljević, G.T. (2004) *Chem. Ing. Tech.*, **76**, 376–383.
42. Gijsbertsen-Abrahamse, A.J., Van der Padt, A., and Boom, R.M. (2004) *J. Membr. Sci.*, **230**, 149–159.
43. Yuan, Q., Williams, R.A., and Aryanti, N. (2010) *Adv. Powder Technol.*, **21**, 599–608.
44. Timgren, A., Trägårdh, G., and Trägårdh, C. (2010) *Chem. Eng. Res. Des.*, **88**, 229–238.
45. De Luca, G., Di Maio, F.P., Di Renzo, A., and Drioli, E. (2008) *Chem. Eng. Process.*, **47**, 1150–1158.
46. Christov, N.C., Danov, K.D., Danova, D.K., and Kralchevsky, P.A. (2008) *Langmuir*, **24**, 1397–1410.
47. Williams, R.A., Peng, S.J., Wheeler, D.A., Morley, N.C., Taylor, D., Whalley, M., and Houldsworth, D.W. (1998) *Chem. Eng. Res. Des.*, **76 A**, 902–910.
48. Nakashima, T. and Shimizu, M. (1986) *Ceram. Jpn.*, **21**, 408–412 (in Japanese).
49. Vladisavljević, G.T., Kobayashi, I., Nakajima, M., Williams, R.A., Shimizu, M., and Nakashima, T. (2007) *J. Membr. Sci.*, **302**, 243–253.
50. Vladisavljević, G.T., Shimizu, M., and Nakashima, T. (2005) *J. Membr. Sci.*, **250**, 69–77.
51. Kukizaki, M. and Nakashima, T. (2004) *Membrane*, **29**, 301–308.
52. Kukizaki, M. and Goto, M. (2007a) *J. Membr. Sci.*, **299**, 190–199.
53. Kukizaki, M. (2010) *J. Membr. Sci.*, **360**, 426–435.
54. Vladisavljević, G.T., Tesch, S., and Schubert, H. (2002a) *Chem. Eng. Process.*, **41**, 231–238.
55. Egidi, E., Gasparini, G., Holdich, R.G., Vladisavljević, G.T., and Kosvintsev, S.R. (2008) *J. Membr. Sci.*, **323**, 414–420.
56. Vladisavljević, G.T. and Schubert, H. (2003a) *J. Dispersion Sci. Technol.*, **24**, 811–819.
57. Van der Graaf, S., Schroën, C.G.P.H., Van der Sman, R.G.M., and Boom, R.M. (2004) *J. Colloid Interface Sci.*, **277**, 456–463.
58. Rayner, M., Trägårdh, G., and Trägårdh, C. (2005) *Colloids Surf. A*, **266**, 1–17.

59. Nakashima, T., Shimizu, M., and Kukizaki, M. (1993) *Kag. Kog. Ronbunshu*, **19**, 991–997 (in Japanese).
60. Surh, J., Jeong, Y.G., and Vladisavljević, G.T. (2008) *J. Food Eng.*, **89**, 164–170.
61. Vladisavljević, G.T. and McClements, D.J. (2010) *Colloid Surf. A*, **364**, 123–131.
62. Whitesides, G.M. (2006) *Nature*, **442**, 368–373.
63. Kikuchi, Y., Ohki, H., Kaneko, T., and Sato, K. (1989) *Biorheology*, **26**, 1055. (abstr.)
64. Kikuchi, Y., Sate, K., Ohki, H., and Kaneko, T. (1992) *Microvasc. Res.*, **44**, 226–240.
65. Manz, A., Harrison, D.J., Verpoorte, E.D.J., Fettinger, J.C., Paulus, A., Lüdi, H., and Widmer, H.M. (1992) *J. Chromatogr.*, **593**, 253–258.
66. Atencia, J. and Beebe, D.J. (2005) *Nature*, **437**, 648–655.
67. Kameoka, J., Craighead, H.G., Zhang, H.W., and Henion, J. (2001) *Anal. Chem.*, **73**, 1935–1941.
68. Hatch, A., Kamholz, A.E., Hawkins, K.R., Munson, M.S., Schilling, E.A., Weigl, B.H., and Yager, P. (2001) *Nat. Biotechnol.*, **19**, 461–465.
69. Andersson, H. and van de Berg, A. (2003) *Sens. Actuators B*, **92**, 315–325.
70. Figeys, D. and Pinto, D. (2001) *Electrophoresis*, **22**, 208–216.
71. McClain, M.A., Culbertson, C.T., Jacobson, S.C., and Ramsey, J.M. (2001) *Anal. Chem.*, **73**, 5334–5338.
72. Thorsen, T., Roberts, R.W., Arnold, F.H., and Quake, S.R. (2001) *Phys. Rev. Lett.*, **86**, 4163–4166.
73. Link, D.R., Anna, S.L., Weitz, D.A., and Stone, H.A. (2004) *Phys. Rev. Lett.*, **92** (Art. No.: 054503), 1–4.
74. Yi, G.R., Thorsen, T., Manoharan, V.N., Hwang, M.J., Jeon, S.J., Pine, D.J., Quake, S.R., and Yang, S.M. (2003) *Adv. Mater.*, **15**, 1300–1304.
75. Tice, J.D., Lyon, A.D., and Ismagilov, R.F. (2004) *Anal. Chim. Acta*, **507**, 73–77.
76. He, M., Edgar, J.S., Jeffries, G.D.M., Lorenz, R.M., Shelby, J.P., and Chiu, D.T. (2005) *Anal. Chem.*, **77**, 1539–1544.
77. Van der Graaf, S., Steegmans, M.L.J., Van der Sman, R.G.M., Schroën, C.G.P.H., and Boom, R.M. (2005) *Colloid Surf. A*, **266**, 106–116.
78. Dendukuri, D., Tsoi, K., Hatton, T.A., and Doyle, P.S. (2005) *Langmuir*, **21**, 2113–2116.
79. Nisisako, T., Torii, T., and Higuchi, T. (2004) *Chem. Eng. J.*, **101**, 23–29.
80. Tan, Y.C., Cristini, V., and Lee, A.P. (2006) *Sens. Actuator B-Chem.*, **114**, 350–356.
81. Steegmans, M.L.J., Schroën, K.G.P.H., and Boom, R.M. (2009) *Langmuir*, **25**, 3396–3401.
82. Xia, Y. and Whitesides, G.M. (1998) *Annu. Rev. Mater. Sci.*, **28**, 153–184.
83. Nie, Z., Xu, S., Seo, M., Lewis, P.C., and Kumacheva, E. (2005) *J. Am. Chem. Soc.*, **127**, 8058–8063.
84. Zhao, C.X. and Middelberg, A.P.J. (2011) *Chem. Eng. Sci.*, **66**, 1394–1411.
85. Okushima, S., Nisisako, T., Torii, T., and Higuchi, T. (2004) *Langmuir*, **20**, 9905–9908.
86. Nisisako, T., Okushima, S., and Torii, T. (2005) *Soft Matter*, **1**, 23–27.
87. Tan, Y.C., Fisher, J.S., Lee, A.I., Cristini, V., and Lee, A.P. (2004) *Lab Chip*, **4**, 292–298.
88. Choi, C.H., Jung, J.H., Rhee, Y.W., Kim, D.P., Shim, S.E., and Lee, C.S. (2007) *Biomed. Microdevices*, **9**, 855–862.
89. Rayleigh, L. (1879) *Proc. Lond. Math. Soc.*, **10**, 4–13.
90. Abate, A.R., Thiele, J., and Weitz, D.A. (2011) *Lab Chip*, **11**, 253–258.
91. Abate, A.R. and Weitz, D.A. (2009) *Small*, **5**, 2030–2032.
92. Shepherd, R.F., Conrad, J.C., Rhodes, S.K., Link, D.R., Marquez, M., Weitz, D.A., and Lewis, J.A. (2006) *Langmuir*, **22**, 8618–8622.
93. Yi, G.R., Thorsenxx, T., Garstecki, P., Stone, H.A., and Whitesides, G.M. (2005) *Phys. Rev. Lett.*, **94** (Art. No.: 164501), 1–4.
94. Lewis, P.C., Graham, R.R., Nie, Z., Xu, S., Seo, M., and Kumacheva, E. (2005) *Macromolecules*, **38**, 4536–4538.
95. Xu, Q. and Nakajima, M. (2004) *Appl. Phys. Lett.*, **85**, 3726–3728.

96. Anna, S.L., Bontoux, N., and Stone, H.A. (2003) *Appl. Phys. Lett.*, **82**, 364–366.
97. Seo, M., Paquet, C., Nie, Z., Xu, S., and Kumacheva, E. (2007) *Soft Matter*, **3**, 986–992.
98. Yobas, L., Martens, S., Ong, W.L., and Ranganathan, N. (2006) *Lab Chip*, **6**, 1073–1079.
99. Kobayashi, I., Mukataka, S., and Nakajima, M. (2005a) *Ind. Eng. Chem. Res.*, **44**, 5852–5856.
100. Kobayashi, I., Nakajima, M., Chun, K., Kikuchi, Y., and Fujita, H. (2002) *AIChE J.*, **48**, 1639–1644.
101. Tong, J., Nakajima, M., Nabetani, H., Kikuchi, Y., and Maruta, Y. (2001) *J. Colloid Interface Sci.*, **237**, 239–248.
102. Liu, H., Nakajima, M., Nishi, T., and Kimura, T. (2005) *Eur. J. Lipid Sci. Technol.*, **107**, 481–487.
103. Sugiura, S., Nakajima, M., Kumazawa, N., Iwamoto, S., and Seki, M. (2002a) *J. Phys. Chem. B*, **106**, 9405–9409.
104. Sugiura, S., Nakajima, M., and Seki, M. (2002b) *Langmuir*, **18**, 5708–5712.
105. Nakagawa, K., Iwamoto, S., Nakajima, M., Shono, A., and Satoh, K. (2004) *J. Colloid Interface Sci.*, **278**, 198–205.
106. Sugiura, S., Nakajima, M., Itou, H., and Seki, M. (2001) *Macromol. Rapid Commun.*, **22**, 773–778.
107. Sugiura, S., Nakajima, M., Tong, J., Nabetani, H., and Seki, M. (2000) *J. Colloid Interface Sci.*, **227**, 95–103.
108. Kawakatsu, T., Komori, H., Nakajima, M., Kikuchi, Y., Komori, H., and Yonemoto, Y. (1999) *J. Chem. Eng. Jpn.*, **32**, 241–244.
109. Kawakatsu, T., Trägårdh, G., Kikuchi, Y., Nakajima, M., Komori, H., and Yonemoto, T. (2000) *J. Surfactants Deterg.*, **3**, 295–302.
110. Kobayashi, I., Wada, Y., Uemura, K., and Nakajima, M. (2010) *Microfluid. Nanofluid.*, **8**, 255–262.
111. Sugiura, S., Nakajima, M., and Seki, M. (2002c) *Langmuir*, **18**, 3854–3859.
112. Kobayashi, I., Mukataka, S., and Nakajima, M. (2004) *J. Colloid Interface Sci.*, **279**, 277–280.
113. van Dijke, K., Kobayashi, I., Schroën, K., Uemura, K., Nakajima, M., and Boom, R. (2010) *Microfluid. Nanofluid.*, **9**, 77–85.
114. Kobayashi, I., Mukataka, S., and Nakajima, M. (2005b) *Langmuir*, **21**, 7629–7632.
115. Kobayashi, I., Wada, Y., Uemura, K., and Nakajima, M. (2009) *Microfluid. Nanofluid.*, **7**, 107–119.
116. Neves, M.A., Ribeiro, H.S., Fujiu, K.B., Kobayashi, I., and Nakajima, M. (2008) *Ind. Eng. Chem. Res.*, **47**, 6405–6411.
117. Vladisavljević, G.T., Kobayashi, I., and Nakajima, M. (2008) *Powder Technol.*, **183**, 37–45.
118. Vladisavljević, G.T., Kobayashi, I., and Nakajima, M. (2011a) *Microfluid. Nanofluid.*, **10**, 1199–1209.
119. Kobayashi, I., Takayuki, T., Maeda, R., Wada, Y., Uemura, K., and Nakajima, M. (2008a) *Microfluid. Nanofluid.*, **4**, 167–177.
120. Kobayashi, I., Wada, Y., Uemura, K., and Nakajima, M. (2008b) *Microfluid. Nanofluid.*, **5**, 677–687.
121. Sugiura, S., Oda, T., Izumida, Y., Aoyagi, Y., Satake, M., Ochiali, A., Ohkohchi, N., and Nakajima, M. (2005) *Biomaterials*, **26**, 3327–3331.
122. Utada, A.S., Lorenceau, E., Link, D.R., Kaplan, P.D., Stone, H.A., and Weitz, D.A. (2005) *Science*, **308**, 537–541.
123. Utada, A.S., Chu, L.-Y., Fernandez-Nieves, A., Link, D.R., Holtze, C., and Weitz, D.A. (2007) *MRS Bull.*, **32**, 702–708.
124. Utada, A.S., Fernandez-Nieves, A., Gordillo, J.M., and Weitz, D.A. (2008) *Phys. Rev. Lett.*, **100** (Art. No.: 014502), 1–4.
125. Shah, R.K., Shum, H.C., Rowat, A.C., Lee, D., Agresti, J.J., Utada, A.S., Chu, L.Y., Kim, J.W., Fernandez-Nieves, A., Martinez, C.J., and Weitz, D.A. (2008) *Mater. Today*, **11**, 18–27.
126. Chu, L.Y., Utada, A.S., Shah, R.K., Kim, J.W., and Weitz, D.A. (2007) *Angew. Chem. Int. Ed.*, **119**, 9128–9132.
127. Vladisavljević, G.T., Duncanson, W.J., Shum, H.C., and Weitz, A.D. (2011b) Fabrication of biodegradable poly(lactic acid) particles in flow focusing glass capillary devices. UK Colloids 2011, London.

128. Kim, J.W., Utada, A.S., Alberto Fernández-Nieves, A., Hu, Z., and Weitz, D.A. (2007) *Angew. Chem. Int. Ed.*, **46**, 1819–1822.
129. Shum, H.C., Lee, D., Yoon, I., Kodger, T., and Weitz, D.A. (2008) *Langmuir*, **24**, 7651–7653.
130. Lorenceau, E., Utada, A.S., Link, D.R., Cristobal, G., Joanicot, M., and Weitz, D.A. (2005) *Langmuir*, **21**, 9183–9186.
131. Lee, D. and Weitz, D.A. (2008) *Adv. Mater.*, **20**, 3498–3503.
132. Vladisavljević, G.T., Shum, H.C., and Weitz, A.D. (2011c) Control over the shell thickness of core/shell drops in three-phase glass capillary devices. UK Colloids 2011, London.
133. Liu, X.D., Bao, D.C., Xue, W.M., Xiong, Y., Yu, W.T., Yu, X.J., Ma, X.J., and Yuan, Q. (2003) *J. Appl. Polym. Sci.*, **87**, 848–852.
134. Ma, G.H., Nagai, M., and Omi, S. (1999) *Colloid Surf. A*, **153**, 383–394.
135. Yanagishita, T., Tomabechi, Y., Nishio, K., and Masuda, H. (2004) *Langmuir*, **20**, 554–555.
136. Yamamoto, T., Ohmori, T., and Kim, Y.H. (2010) *Carbon*, **48**, 912–928.
137. Kakazu, E., Murakami, T., Akamatsu, K., Sugawara, T., Kikuchi, R., and Nakao, S. (2010) *J. Membr. Sci.*, **354**, 1–5.
138. Kukizaki, M. and Goto, M. (2007b) *Colloid Surf. A*, **293**, 87–94.
139. Sawalha, H., Fan, Y., Schroën, K., and Boom, R. (2008) *J. Membr. Sci.*, **325**, 665–671.
140. Chu, L.Y., Xie, R., Zhu, J.H., Chen, W.M., Yamaguchi, T., and Nakao, S. (2003) *J. Colloid Interface Sci.*, **265**, 187–196.
141. Shum, H.C., Abate, A.R., Lee, D., Studart, A.R., Wang, B., Chen, C.H., Thiele, J., Shah, R.K., Krummel, A., and Weitz, D.A. (2010) *Macromol. Rapid Commun.*, **31**, 108–118.
142. Brouzes, E., Medkova, M., Savaneli, N., Marran, D., Twardowski, M., Hutchison, J.B., Rothberg, M., Link, D.R., Perrimon, N., and Samuels, M.L. (2009) *Proc. Natl. Acad. Sci. U.S.A.*, **106**, 14195–14200.

3
Adsorption Characteristics of Ionic Surfactants at Water/Hexane Interface Obtained by PAT and ODBA

Nenad Mucic, Vincent Pradines, Aliyar Javadi, Altynay Sharipova, Jürgen Krägel, Martin E. Leser, Eugene V. Aksenenko, Valentin B. Fainerman, and Reinhard Miller

3.1
Introduction

Surfactant behavior in bulk solution or at liquid/fluid interfaces has been investigated worldwide since many years by many scientific groups because of their numerous applications in food processing, pharmaceuticals, or cosmetics [1–4]. In contrast to the huge number of papers characterizing the water/air interface [5–10], only few systematic experimental studies and deep theoretical treatment of the data have been dedicated to the surfactant adsorption at water/oil interface [11–15]. Since many years ago, several fundamental technical problems with the oil phase were the main drawback to systematic investigations of adsorption layers at water/oil interfaces. These problems were usually (i) the decrease of sensibility of optical methods owing to the presence of the upper oil phase; (ii) the purification of the oil phases, which often contain traces of highly surface active molecules; (iii) the small density difference of many oils to water, which makes the application of weight-based methods difficult; and (iv) the volatility of the organic solvent, which restricts them to short experimental times owing to its evaporation. By the introduction of more powerful single drop methods, such as the drop profile and capillary pressure techniques, the technical drawbacks for studies of water/oil interface were overcome.

3.2
Experimental Tools

The experimental tools presented here are the profile analysis tensiometer (PAT-1) and oscillating drop and bubble pressure analyzer (ODBA). The ODBA is an experimental module that represents an extension of the standard PAT-1. Both instruments used in this study are products of SINTERFACE Technologies Berlin, Germany [16].

Emulsion Formation and Stability, First Edition. Edited by Tharwat F. Tadros.
© 2013 Wiley-VCH Verlag GmbH & Co. KGaA. Published 2013 by Wiley-VCH Verlag GmbH & Co. KGaA.

PAT allows measuring the surface/interfacial tension of liquids via the analysis of the shape of a buoyant bubble or a pendant drop: in both cases, attached to the tip of a capillary. The method is suitable for liquid/gas (water/air) and liquid/liquid (water/oil) interfaces, and applicable to systems ranging from molten metals to pure organic solvents and diluted and concentrated solutions. There is also no limitation on the magnitude of surface or interfacial tension, accessible in a broad range of temperatures and external pressures [17]. For measurements at constant interfacial area, the time window ranges from about 1 s up to hours and days, so that even extremely slow processes can be followed. The concept of the instrumentation looks rather simple and is shown schematically in Figure 3.1. In particular, the PAT-1 consists mainly of a dosing system driven by computer software, a measuring chamber, and a CCD camera. The dosing system produces a drop/bubble of the test liquid at the tip of an appropriate capillary. The capillary should provide a good wetting such that axisymmetric drops/bubbles are formed. Usually, metal and plastic capillaries are used, depending on the type of solvents and surfactants studied. Another important part of the PAT is the measuring chamber with a glass cuvette in which the respective capillary sits. The glass cuvette contains the second immiscible liquid or gas (air). For drops in air, the cuvette contains some liquid at the bottom, in order to provide a saturated atmosphere for minimizing evaporation effects from the drop. The drop profile tensiometry is applicable also for studies of the dilational viscoelasticity at low frequencies, as was shown for the first time by Benjamins *et al.* [18].

The ODBA uses several elements of PAT (Figure 3.2); however, it provides important additional information, which by no other means is experimentally available. One of the benefits is the capability for measurements at short adsorption times, for example, below 1 s [19]. In addition, this method shows good results for interfaces between two liquids that have a similar density, as it is the usual case under microgravity conditions [20]. This also holds for liquid/liquid systems when both liquids have a similar density. The drawbacks of PAT, such as the inability to measure small spherical drops/bubbles, are compensated by the ODBA technique. This method also allows looking into emulsion films to study directly the film tension or the film elasticity, as it was proposed, for example, by Soos *et al.* [21] and others [22, 23]. An additional important feature of the ODBA is the

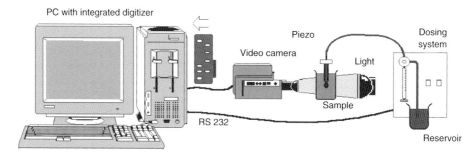

Figure 3.1 Schematic of a drop and bubble profile analysis tensiometer (PAT).

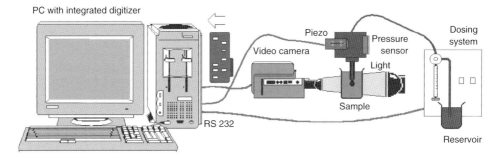

Figure 3.2 Schematic of an oscillating drop and bubble analyzer (ODBA).

option to generate high-frequency oscillations of small spherical drops/bubbles for determining the viscoelasticity of interfacial layers over a broad frequency range.

So far, there are still open questions concerning the importance of the molecular surfactant–surfactant and surfactant–solvent interactions and also concerning the role of oil molecules at the interface, which are responsible for the differences observed when the properties of water/air and water/oil interfaces are compared. The following paragraphs are concerned with the description and quantitative analysis of the adsorption of a series of alkyltrimethylammonium bromides of different alkyl chain lengths (10, 12, 14, 16 carbon atoms) at the water/hexane interface. For the sake of simplicity, we use the abbreviation C_nTAB instead of the conventional nomenclature of these surfactants (DeTAB, DoTAB, TTAB, and CTAB), where n corresponds to the number of carbon atoms in the hydrophobic chain. Interfacial tension isotherms have been measured and interpreted by two different adsorption models giving access to the thermodynamic adsorption parameters. The salt effects due to the presence of phosphate buffer, typically used in single protein or protein–surfactant mixed solutions to stabilize the pH of the aqueous phase, have been taken into account using a Frumkin ionic compressibility model (FIC). On the other side, this model is compared with the basic Frumkin compressibility model (FC) [24].

3.3
Theory

The adsorption behavior of most nonionic and ionic surfactants is very well described by the Frumkin model taking into account the additional lateral interactions between adsorbed surfactant molecules at the interface. In the framework of an electroneutral surface layer, we get the following FIC adsorption model [24]

$$\Pi = -\frac{2RT}{\omega}\left[\ln(1-\theta) + a\theta^2\right] \quad (3.1)$$

$$b[c(c+c_2)]^{1/2}f = \frac{\theta}{1-\theta}\exp(-2a\theta) \quad (3.2)$$

Here, $\theta = \omega \times \Gamma$ is the surface coverage, Γ is the surface concentration, b is the adsorption equilibrium constant, a is the Frumkin parameter corresponding to the interaction constant, R and T are gas law constant and absolute temperature, f is the average activity coefficient of ions in the bulk solution, c is the ionic surfactant concentration, and c_2 is the inorganic (1 : 1) salt concentration. The presence of the prefactor 2 in Eq. (3.1) takes into account the area of the counterion. For short-range interactions, the Debye–Hückel equation corrects accurately the values of the average activity coefficient f:

$$\log f = -\frac{0.5115 \sqrt{I}}{1 + 1.316 \sqrt{I}} + 0.055\, I \tag{3.3}$$

where $I = c + c_2$ is the ionic strength expressed in moles per liter. The given numerical constants correspond to a temperature of 25 °C [25].

The molar area ω can be constant or dependent on Π, as it was shown, for example, in [26]

$$\omega = \omega_0 (1 - \varepsilon \Pi \theta) \tag{3.4}$$

where ω_0 is the molar area at $\Pi = 0$ and ε is the two-dimensional relative surface layer compressibility coefficient. This parameter ε characterizes the intrinsic compressibility of the molecules in the surface layer and can be physically understood, for example, by a change in the tilt angle of adsorbed molecules with increasing surface coverage [27].

The classical Frumkin model [28] also used in this study is given by Eqs. (3.5) and (3.6):

$$\Pi = -\frac{RT}{\omega} \left[\ln(1 - \theta) + a\theta^2 \right] \tag{3.5}$$

$$bc = \frac{\theta}{1 - \theta} \exp(-2a\theta) \tag{3.6}$$

3.4
Results

Interfacial tension measurements have been performed for four members of the homologous series of the cationic surfactants C_nTAB ($n = 10, 12, 14, 16$) at the water/hexane interfaces in the presence of phosphate buffer (10 mM, pH 7). As we have used ionic surfactants with a negligible solubility in hexane, any transfer of surfactant into the oil phase is negligible and therefore a partitioning equilibrium constant totally in favor of the aqueous phase can be assumed. The experimental isotherms, given in Figure 3.3, were fitted using the above presented Frumkin adsorption models and a constant compressibility coefficient of $\varepsilon = 0.01$ mN m^{-1}. This value will not be commented in detail here because its influence is mainly visible on the dilational rheological data, the target of a forthcoming work. The other model parameters obtained from fitting are summarized in Table 3.1.

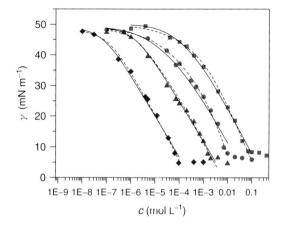

Figure 3.3 Interfacial tension isotherms of C_{10}TAB (■), C_{12}TAB (●), C_{14}TAB (▲), and C_{16}TAB (♦) at the water/hexane interface in presence of phosphate buffer (10 mM, pH 7); solid line corresponds to the Frumkin ionic compressibility model; dashed line corresponds to the Frumkin compressibility model.

Table 3.1 Parameters of adsorption isotherms obtained with the Frumkin ionic compressibility model (salt concentration considered: 2×10^{-3} mol l^{-1}) and basic Frumkin compressibility model at the water/hexane interface [24].

CnTAB	ω_0 (10^5 m^2 mol^{-1})		a		b (10^3 l mol^{-1})	
	FIC	FC	FIC	FC	FIC	FC
$n = 10$	5.6	4.86	0	−0.54	2.2	49.9
$n = 12$	4.4	5.5	0	−1.76	4.3	2.2×10^3
$n = 14$	4.4	4.3	1.2	0	5.5	6.9×10^3
$n = 16$	3.8	4.2	1.1	0	22.0	1.5×10^4

From the thermodynamic point of view, at the water/hexane interface the adsorption starts at much lower concentrations as compared to the water/air interface [24]: for C_{14}TAB and C_{16}TAB, four decades are now necessary to reach the critical micelle concentration (CMC) and this effect is even more pronounced for C_{10}TAB and C_{12}TAB where the isotherm spans over almost six decades of concentration. Owing to the affinity of the hydrophobic chains to the oil phase, the surfactants interact with the alkane molecules located at the interface that favors their adsorption [29]. In Table 3.1, the FIC model gives most reliable results that are easily confirmed by physics. On the other side, the surfactant interaction parameter a has a negative value when obtained from the FC model, which does not have a clear physical meaning. Therefore, for explanation of the thermodynamic properties, the FIC model is more suitable. For analyzing the dynamic properties of the interfacial layers, an interpretation of the FC model is still necessary because no adequate fitting program for the FIC model exists presently. The surfactant interaction parameter $a = 0$ for the FIC model points to the fact that solvent molecules intercalate into the surfactant layer decreasing the mutual surfactant interactions. For C_nTAB with alkyl chains $n \geq 14$, the molecules have mutual interactions strong enough to replace the intercalated hexane molecules at the interface, reflected by the given values of a.

Referring to the work of Mucic et al. [30], it is easy to conclude that for investigations of the kinetics and dilational rheology of C_nTABs we might need to involve the ODBA technique. The Figures 3.4–3.6 are produced following the same procedure given in the respective paper, but in this case the calculations are based on the FC model. As shown in Figure 3.3, the FIC and FC models have a minor difference in the adsorption isotherm shape. Therefore, for further dynamics investigation, the FC model is sufficiently reliable.

In Figure 3.4, an example of the adsorption kinetics of C_{10}TAB obtained at small concentrations is shown, where the interface tension decreases no more than 2mN m^{-1}, at a high solution concentration around the CMC. The kinetics of middle concentrations is placed between these two curves in the graph. From all investigated surfactants in this work, C_{10}TAB is the surfactant with the fastest kinetics. On the other side, the measuring time ranges of ODBA and PAT are 0.01 s to hours and 1 s to hours, respectively. Reflecting these values to Figure 3.4 it is rather simple to conclude how much benefit can be gained involving ODBA

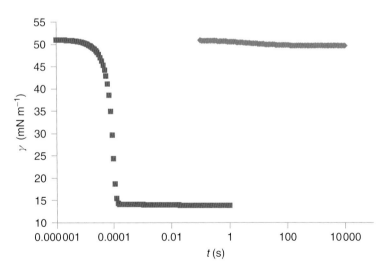

Figure 3.4 Adsorption kinetics of C_{10}TAB calculated from the Ward and Tordai equation are based on the Frumkin compressibility model at the water/hexane interface: ♦ presents the bulk concentration at which the equilibrium surface tension is decreased by approximately 2mN m^{-1} (7.5 × 10^{-6} mol l^{-1}); ■ presents the CMC value, 5 × 10^{-2} mol l^{-1}.

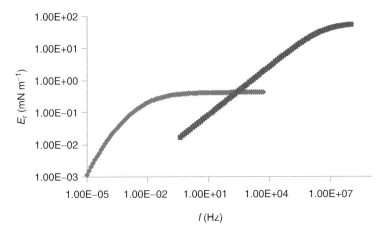

Figure 3.5 The real part of the dilational viscoelasticity for $C_{10}TAB$: ♦ presents the bulk concentration where surface tension equilibrium is decreased by around 2 mN m^{-1} (7.5 × 10^{-6} mol l^{-1}); ■ presents the CMC value, 5 × 10^{-2} mol l^{-1}.

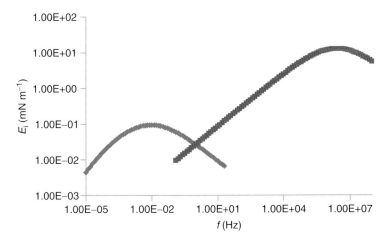

Figure 3.6 The imaginary part of the dilational viscoelasticity of $C_{10}TAB$ solutions: ♦ presents the bulk concentration where around 2 mN m^{-1} surface tension equilibrium is decreased (7.5 × 10^{-6} mol l^{-1}); ■ presents the CMC value, 5 × 10^{-2} mol l^{-1}.

in $C_{10}TAB$ dynamic investigation. The same situation is true for the rheology investigation.

Figures 3.5 and 3.6 show the dilational elasticity and viscosity of $C_{10}TAB$ calculated using the freely available IsoFit program [31]. The measuring frequency ranges for ODBA and PAT are 0.1–100 Hz and 0.001–0.1 Hz, respectively. Therefore, these two techniques gathered together can cover the necessary frequency range for reliable investigations of the viscoelasticity of $C_{10}TAB$ solutions. The same conclusion can be drawn for the other studied C_nTAB molecules where PAT technique is not sufficiently fast for the dynamic investigation.

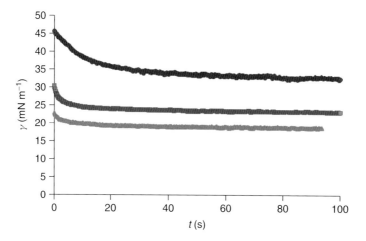

Figure 3.7 Adsorption kinetics of $C_{16}TAB$ pure aqueous solution obtained by ODBA: (●) $c = 2 \times 10^{-5}$ mol l^{-1}; (■) $c = 10^{-4}$ mol l^{-1}, and (▲) $c = 2 \times 10^{-4}$ mol l^{-1}.

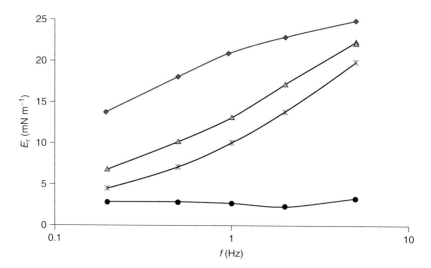

Figure 3.8 The real part of the $C_{16}TAB$ pure aqueous solution viscoelasticity obtained by ODBA: (◆) $c = 2 \times 10^{-5}$ mol l^{-1}, (▲) $c = 10^{-4}$ mol l^{-1}, (∗) $c = 2 \times 10^{-4}$ mol l^{-1}, and (●) pure water.

As an example of practical application of the ODBA, the adsorption kinetics and the real part of the dilational viscoelasticity for $C_{16}TAB$ solutions are represented in Figures 3.7 and 3.8. All solutions are prepared in pure milli-Q water that makes them different from the solutions in Figure 3.3, which were prepared in a phosphate buffer of pH 7. Therefore, in this situation, the screening effect of dissolved electrolytes is omitted and the adsorption isotherm is shifted toward higher bulk concentrations. At the same time, the adsorption coefficient b decreases.

In the given example, only part of the full capacity of the ODBA technique is presented. Thus, in Figures 3.7 and 3.8, the adsorption time starts from 0.1 s and the maximum frequency is 5 Hz, respectively. At higher frequencies, the viscoelasticity can be obtained only after a careful quantitative consideration of the hydrodynamic effects occurring in and around the drop. More about the features of the ODBA and its fields of application are summarized in [32].

3.5
Summary

In summary, it can be said that this work presents a qualitative and quantitative analysis of interface properties of C_nTABs adsorption layers at the water/hexane interface. Beside that, a comparison of appropriate adsorption models took place. It was shown that for the qualitative analysis FIC model was used. This model reveals that at high surface coverage hexane molecules remain in the C_{10}TAB and C_{12}TAB adsorption layers between packed surfactants. On the other side, C_{14}TAB and C_{16}TAB have strong mutual interactions that do not allow hexane molecules to penetrate into the adsorption layer. For quantitative analysis, the FC model is sufficiently capable of giving a clue as to what instrument would be suitable for investigation of adsorption dynamics. For this purpose, ODBA technique is sufficiently fast to gather together with PAT, as a combination for adsorption kinetics and dilational rheology investigation.

Acknowledgments

The work was financially supported by projects of ESA (FASES MAP AO-99-052), the German Space Agency (DLR 50WM0941), and the DFG SPP 1506 (Mi418/18-1).

References

1. Schramm, L.L., Stasiuk, E.N., and Marangoni, D.G. (2003) *Annu. Rep. Prog. Chem.*, **99**, 3–48.
2. Müller, R.H. and Keck, C.M. (2004) *J. Biotechnol.*, **113**, 151–170.
3. Leser, M.E., Sagalowicz, L., Michel, M., and Watzke, H.J. (2006) *Adv. Colloid Interface Sci.*, **123–126**, 125–136.
4. Dickinson, E. (2008) *Soft Mater.*, **4**, 932–942.
5. Chang, C.-H. and Franses, E.I. (1995) *Colloids Surf., A*, **100**, 1–45.
6. Warszynski, P., Barzyk, W., Lunkenheimer, K., and Fruhner, H. (1998) *J. Phys. Chem. B*, **102**, 10948–10957.
7. Fainerman, V.B., Miller, R., Aksenenko, E.V., Makievski, A.V., Krägel, J., Loglio, G., and Liggieri, L. (2000) *Adv. Colloid Interface Sci.*, **86**, 83–101.
8. Fainerman, V.B., Miller, R., and Aksenenko, E.V. (2002) *Adv. Colloid Interface Sci.*, **96**, 339–359.
9. Kolev, V.L., Danov, K.D., Kralchevsky, P.A., Broze, G., and Mehreteab, A. (2002) *Langmuir*, **18**, 9106–9109.
10. Pradines, V., Lavabre, D., Micheau, J.-C., and Pimienta, V. (2005) *Langmuir*, **21**, 11167–11172.

11. Deshiikan, S.R., Bush, D., Eschenazi, E., and Papadopoulos, K.D. (1998) *Colloids Surf., A*, **136**, 133–150.
12. Mulqueen, M. and Blankschtein, D. (2001) *Langmuir*, **18**, 365–376.
13. Zarbakhsh, A., Querol, A., Bowers, J., Yaseen, M., Lu, J.R., and Webster, J.R.P. (2005) *Langmuir*, **21**, 11704–11709.
14. Pradines, V., Krägel, J., Fainerman, V.B., and Miller, R. (2009) *J. Phys. Chem. B*, **113**, 745–751.
15. Kotsmar, C., Aksenenko, E.V., Fainerman, V.B., Pradines, V., and Miller, R. (2010) *Colloids Surf., A*, **354**, 210–217.
16. (2012) See for example www.sinterface.com.
17. Neumann, A.W. and Spelt, J.K. (1996) in *Applied Surface Thermodynamics* (eds J.K. Spelt and A.W. Neumann), Marcel Dekker, New York, pp. 333–378.
18. Benjamins, J., Cagna, A., and Lucassen-Reynders, E.H. (1996) *Colloids Surf., A*, **114**, 245–254.
19. Javadi, A., Krägel, J., Pandolfini, P., Loglio, G., Kovalchuk, V.I., Aksenenko, E.V., Ravera, F., Liggieri, L., and Miller, R. (2010) *Colloids Surf., A*, **365**, 62.
20. Kovalchuk, V.I., Ravera, F., Liggieri, L., Loglio, G., Pandolfini, P., Makievski, A.V., Vincent-Bonnieu, S., Krägel, J., Javadi, A., and Miller, R. (2010) *Adv. Colloid Interface Sci.*, **161**, 102.
21. Soos, J.M., Koczo, K., Erdos, E., and Wasan, D.T. (1994) *Rev. Sci. Instrum.*, **65**, 3555.
22. Kovalchuk, V.I., Makievski, A.V., Krägel, J., Pandolfini, P., Loglio, G., Liggieri, L., Ravera, F., and Miller, R. (2005) *Colloids Surf., A*, **261**, 115–121.
23. Georgieva, D., Cagna, A., and Langevin, D. (2009) *Soft Matter*, **5**, 2063–2071.
24. Pradines, V., Fainerman, V.B., Aksenenko, E.V., Krägel, J., Mucic, N., and Miller, R. (2010) *Colloids Surf., A*, **371**, 22–28.
25. Fainerman, V.B. and Lucassen-Reynders, E.H. (2002) *Adv. Colloid Interface Sci.*, **96**, 295–323.
26. Fainerman, V.B., Miller, R., and Kovalchuk, V.I. (2003) *J. Phys. Chem. B*, **107**, 6119–6121.
27. Fainerman, V.B. and Vollhardt, D. (2003) *J. Phys. Chem. B*, **107**, 3098–3100.
28. Frumkin, A. (1925) *Z. Phys. Chem. (Leipzig)*, **116**, 466–484.
29. Medrzycka, K. and Zwierzykowski, W. (2000) *J. Colloid Interface Sci.*, **230**, 67–66.
30. Mucic, N., Javadi, A., Kovalchuk, N.M., Aksenenko, E.V., and Miller, R. (2011) *Adv. Colloid Interface Sci.*, **168**, 167–178. doi: 10.1016/j.cis.2011.06.001
31. (2012) http://www.thomascat.info/thomascat/Scientific/AdSo/AdSo.htm.
32. Javadi, A., Krägel, J., Makievski, A.V., Kovalchuk, N.M., Kovalchuk, V.I., Mucic, N., Loglio, G., Pandolfini, P., Karbaschi, M., and Miller, R. (2012) Fast dynamic interfacial tension measurements and dilational rheology of interfacial layers by using the capillary pressure technique, Colloids Surfaces A, **407**, 159–168.

4
Measurement Techniques Applicable to the Investigation of Emulsion Formation during Processing

Nima Niknafs, Robin D. Hancocks, and Ian T. Norton

4.1
Introduction

A wide range of natural and processed foods are emulsion based, and understanding and controlling their functionality requires knowledge of emulsion formation, stability, storage, and how these systems break down on consumption. Controlling the droplet size of the emulsion is one of the means to induce particular functionalities; droplet size influences important emulsion properties, such as stability, color, and shelf life [1]. The final average emulsion droplet size is the result of the dynamic balance between the two subprocesses of droplet breakup and coalescence, which occur simultaneously during emulsification. Despite previous studies, a lack of understanding of the physical behavior of emulsions during processing remains, mostly because of the lack of direct measurement techniques for monitoring the mechanism of emulsification in real time.

In a typical emulsification process, under steady state conditions the droplet size would be the result of the dynamic balance between droplet breakup and coalescence. However, the time scales of the subprocesses in the early stages of emulsification are significantly short. Thus, droplet sizes and droplet size distributions of any process undergo rapid variations, which create a necessity for robust measurement techniques. An overview of the measurement techniques developed for experimental studies of emulsification is shown in Table 4.1. These techniques can generally be divided into three categories: those monitoring droplet size, those that concentrate on droplet breakup, and those that consider coalescence individually.

Several techniques exist for offline measurement of droplet size using samples obtained from emulsification experiments. The reliability of the measurements of the samples depends on the stability of the samples after sampling and before the measurement. Therefore, it is often a problem when precise measurements are required with samples that are not stable under natural conditions. In order to overcome this problem, after sampling, a large amount of stabilizer is added to the solution in order to ensure that the droplet size will not change until the time of the measurement. However, not only may the time from sampling until the addition

Emulsion Formation and Stability, First Edition. Edited by Tharwat F. Tadros.
© 2013 Wiley-VCH Verlag GmbH & Co. KGaA. Published 2013 by Wiley-VCH Verlag GmbH & Co. KGaA.

Table 4.1 Overview of different emulsification characterization techniques.

Property that is measured	Mode of operation	Types of systems	Measurement technique	Remarks	References/devices
Droplet size	Offline	Laser systems	Light scattering techniques	Dispersed phase volume fraction less than 0.05%	Mastersizer 2000®
		Sound systems	Ultrasonic spectroscopy	Dispersed phase volume fraction less than 40%	Ultrasizer®
	Online	Laser systems	Diffraction technique	Dilute emulsion	[3, 4]
			Doppler anemometry	Dilute emulsion, transparent continuous, and dispersed phase	[5, 6]
			Spatial filtering	Dilute emulsion	[7]
			FBRM	The dispersed phase volume fraction as high as 40%	Alopaeus et al. [8]
		Sound systems	Pulse-echo technique	A measurement device should be developed	Tong and Povey [9]
		Direct imaging	Employing stereomicroscope	Intrusive, the measurement is localized on the wall, image analysis, only for emulsions containing droplet size above 10 μm	Pacek et al. [10]
			Employing endoscope	Intrusive, image analysis	Alban et al. [2]
			PVM®	Intrusive, image analysis, only for emulsions containing droplet size above 10 μm, requires at least 3 min measurement time	Lasentech

		Other	Capillary method	Complex apparatus, suitable for processes at dynamic equilibrium, requires calibration	Bae and Tavlarides [11] and Hocq et al. [12]
			NMR pulse sequence method	At early stages of development	Hollingsworth et al. [13]
			Light transmittance	Dilute emulsions, does not produce droplet size distribution, fast data acquisition rate, requires calibration	Hong and Lee [14]
			Light reflectance	Does not produce droplet size distribution, fast data acquisition rate, requires calibration	Niknafs et al. [15]
Droplet breakup	Offline	Similar to droplet size measurement techniques	Similar to droplet size measurement techniques	Processing conditions: at early stages of emulsification, dispersed phase volume fraction less than 1%, high concentration of stabilizers	Vankova et al. [16]
Droplet coalescence	Online	Measuring a change in a surrogate parameter	Induce a coalescence dominant hydrodynamic condition	Does not examine droplet coalescence in the presence of droplet breakup phenomena	Howarth [17]
	Offline		Turbidity measurement	Chemical modification of the dispersed phase, preemulsions should be prepared, depends on the droplet size	Taisne et al. [18]
			Fluorescent probe	Preemulsion should be prepared, suitable for emulsions containing nonviscous dispersed phases	Lobo et al. [19]
			Coloring technique	Preemulsions should be prepared, the technique is limited by optical microscopy limitations	Danner [20]

to the solution be enough to change the droplet size, but also, in some cases, the addition of even small impurities may change the microstructure of the sample [2]. In order to fully investigate and understand the dynamics of the processes during emulsification, there is a need to determine the droplet size during the process, which enables the determination of the exact phenomena occurring in the processing vessel. Therefore, the major effort in this area is focused on the development of techniques that can be used online and during processing.

It should be noted that the techniques employed for droplet size monitoring are also employed in the investigation of droplet breakup alone. The only difference is in the choice of processing conditions. In order to investigate droplet breakup individually, it should be separated from droplet coalescence. This is achieved by either reducing the dispersed phase volume fraction ($\varphi < 2\%$) while there is high emulsifier concentration, or by measuring the droplet size at certain times during processing, when droplet coalescence is minimal, for example, in the initial stages of the process [16, 21]. This report's attention was focused toward the online techniques.

4.2
Online Droplet Size Measurement Techniques

4.2.1
Laser Systems

Measurement techniques that employ laser light to determine the droplet size are divided into four groups depending on the operating principle. Normally, they can be used online and the data is collected rapidly, making these techniques suitable for emulsification processes. However, laser-based techniques do not necessarily generate accurate data and some of the measurement techniques need additional calibration.

One of the first and widely used techniques to determine the droplet size employs the diffraction of the propagated laser beam [3, 4]. This technique is based on the measurement of diffracted light from droplets passing through the laser beam. The diffracted light is monitored by using a concentric annular probe so that each detector measures light scattered at any angle. Finally, these measurements are used to determine the droplet size distribution by employing Fraunhofer diffraction theory. The advantage of this technique is that it is fast and does not require any additional calibration. However, its main drawback is its limited applicability, as it can be used only for dilute emulsions.

In another approach, the phase Doppler anemometry (PDA) principle is employed for droplet size measurement [5, 6]. This technique measures the phase difference between two scattered light signals at two collection points to determine the droplet size. This technique can only be used for systems where not only the continuous and dispersed phases are transparent, but the entire emulsion is also transparent. Therefore, PDA has little application in this framework.

4.2 Online Droplet Size Measurement Techniques

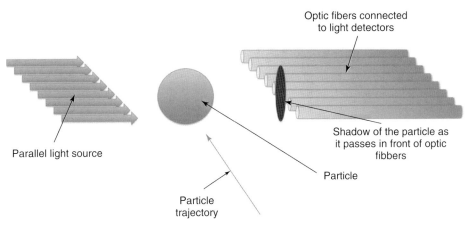

Figure 4.1 The shadow created as the droplet passes between a light source and an array of small size optic fibers. The particle obscures the light reaching the optic fibers as it passes creating a signal, which is detected to provide not only the droplet size but also its velocity.

The third approach [7] uses modified spatial filtering velocimetry (SFV) by fiber-optical spot scanning (FSS) to measure droplet size. SFV is a method of measuring the velocity of a particle by observing it through a filter in front of a receiver. FSS is employed to observe the shadow image of a particle through an optical fiber array with small fiber sizes as shown in Figure 4.1. This results in the generation of impulses detected by the optic fibers, the width of which depends on the droplet size and speed. The advantage of this technique is that it measures the droplet sizes and their velocity, which can be useful in emulsification studies. However, it can only be employed for low dispersed phase volume fractions [22].

One of the most promising techniques using lasers is focused beam reflectance measurement (FBRM) [8]. An infrared laser beam is passed through a prism rotated at a known velocity, propagating the laser beam through a lens on the probe tip. When the beam hits a droplet, the light is backscattered toward the probe window as illustrated in Figure 4.2. The backscattered light is subsequently measured by a detector mounted behind the lens. As the scanning velocity and the time delay of the backscattered light are known, the characteristic length is recorded. Given that the probability of the propagated beam hitting any part of the droplet is the same, a chord length is measured and transformed into the droplet size distribution. The main advantage of this technique is its simplicity and the robustness of the hardware required. In addition, it can be employed for emulsions with dispersed phase volume fractions as high as 40% [8, 23]. However, unsatisfactory results are observed when the measured droplet size distribution obtained by this technique is compared with that determined by other techniques [24]. The reason for this is that, as the measured chord length is not always representative of the droplet diameter, various assumptions should be considered in the mathematical transformation of the data to droplet size distribution. Therefore, different mathematical approaches

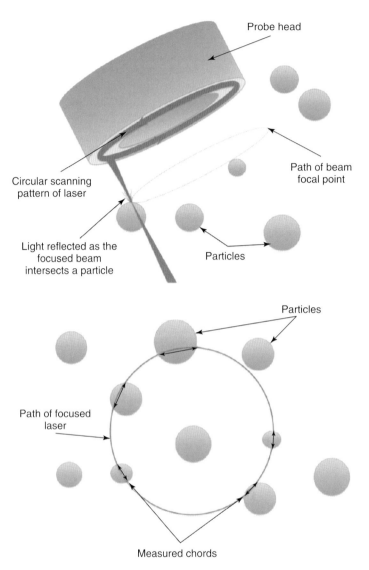

Figure 4.2 Focused beam reflectance measurement uses a beam that is rotated so that it passes over the droplets being measured in a circular motion. The light reflected back from the droplets is collected by the probe and the resultant information is used to determine droplet size distribution. As the path of the beam does not necessarily measure a diameter of each droplet but instead a random chord length, the data requires mathematical transformation after the measurement.

with different levels of sophistication have been developed [23–25], which do not always result in similar droplet sizes. Therefore, although this technique is highly promising, it requires further development.

4.2.2
Sound Systems

Ultrasonic spectroscopy, or acoustic droplet size measurement, has recently attracted considerable attention [26], because of its ability to measure droplet size in opaque systems, such as milk and salad dressing. In addition, wider droplet size ranges can be detected by this technique than by those using laser techniques [27]. Ultrasonic spectroscopy is based on the measurement of the attenuation of ultrasound radiation propagated into an emulsion system. This data is subsequently transformed into droplet size distribution using theoretical models. Various mathematical models have been developed for this reason [27–30]. Additionally, several applications have been reported where ultrasonic spectroscopy is used together with other techniques. An example of such an application is when this technique is employed in conjunction with electrophoresis for the measurement of the electric properties of an emulsion system in addition to the measurement of the droplet size [27].

One of the main drawbacks, that ultrasonic spectroscopy has not found widespread use in industry, is related to the fact that it has an upper limit for the dispersed phase volume fraction of the emulsions in order to measure the droplet size accurately. Although in some cases, 15% dispersed phase volume fraction has been used [9], most of the studies are performed on emulsions with dispersed phase volume fractions smaller than 5% [26, 27, 30]. Nonetheless, the main drawback of this technique is that it is mostly suitable for offline droplet size measurement, although dilution is unnecessary. This is the result of the fact that, in order to determine droplet sizes within a broad range, ultrasonic spectroscopy employs acoustic waves with a wide frequency range (1–100 MHz). Therefore, relatively long time, for example, 45 min [26], in comparison with emulsification time scale, is required for droplet size measurement. It should be noted that a pulse-echo technique was developed for rapid measurement of the droplet size suitable for emulsification processes [9]. This was achieved by reducing the range of sound frequencies and measuring the velocity of the sound instead of its attenuation. However, this approach requires further research and development, in order to become fully operational for online droplet size measurement. Some examples of acoustic measurement cells are shown in Figure 4.3.

4.2.3
Direct Imaging

Direct imaging methods are among the most conventional techniques employed for emulsification characterization. They are the only techniques that can not only

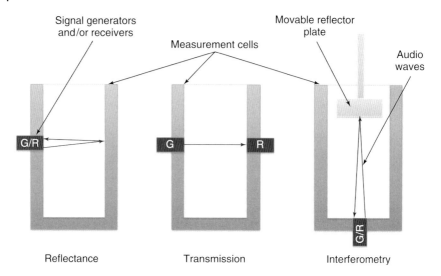

Figure 4.3 The measurement of droplet sizes using ultrasonic waves can be done using several different methods, three of the most common are shown: reflectance, transmission, and interferometry. The sound propagates through the sample in a different path for each technique.

provide the droplet size distribution, but also can determine the shape of the droplets. Three different approaches have been developed employing this concept.

The first online imaging technique was developed by Pacek et al. [10] who employed a video camera attached to a stereomicroscope. This measurement apparatus was set up outside the mixing vessel and in order to obtain clear pictures a stroboscope for illumination was introduced into the mixing vessel. The obtained images were subsequently analyzed with the aid of semiautomatic image analysis software, and the droplet size distribution was determined. Although this technique generates suitable data for determining droplet size distribution in time scales relevant to the emulsification processes, its main drawback is that the images are localized on the wall of the vessel at a depth of up to 8 mm. Moreover, the introduction of a thick stroboscopic light source into the vessel may affect the process. Using this technique, droplet sizes greater than 10 μm can be characterized.

This technique was modified by Galindo et al. [31] who replaced the video recording system with a CCD camera. This increased the depth of the image up to 20 mm. Fully automated image analysis software was developed that employed the Hough transformation. Although the proposed methodology improved the technique, the intrusive nature of this technique still persists because of the use of stroboscopic light source inside the vessel, in addition to the fact that only local images can be obtained.

Another approach using direct imaging was developed by Alban et al. [2]. Using a stereomicroscope combined with a recording system, they employed a stereo probe conventionally used for automated inspection applications. The stereo probe was a monochromic progressive-scan camera. Although the probe had the ability to record videos, it was not possible to obtain a clear image from the fast moving droplets. Therefore, a stroboscopic light source was attached to the stereo probe via an optical fiber. The obtained images were then analyzed employing image analysis software. This methodology eliminates one of the main drawbacks of the technique developed by Pacek et al. [10], enabling images to be freely obtained from various positions from within the mixing vessel. However, the invasive nature of the technique remains, although to a lesser extent. In addition, images were obtained from emulsification in the presence of 70% dispersed volume fractions [2].

A new measurement technique was developed by Lasentech (USA), which is referred to as *particle vision and measurement* (PVM®) [32]. Light from six laser sources is focused to generate a fixed ($2\,\text{mm}^2$) area of illumination as shown in Figure 4.4. When droplets are encountered within this area, they scatter the laser beam in all directions. The backscattered light is measured by a lensing system in the probe and then is relayed on a CCD array. The image captured on the CCD array is analyzed by commercially available image analysis software. Although this technique has been shown to operate better than other commercially available techniques [22], it has been shown that a minimum time of 3 min is required to obtain a sufficient number of images for the accurate determination of a representative droplet size distribution [32]. Therefore, this technique is not suitable for transient emulsification, for example, in the initial stages of the process. In addition, droplet sizes larger than 10 µm have been shown to be the limit of

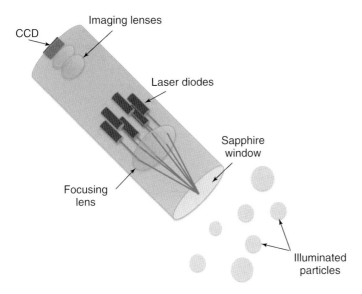

Figure 4.4 The PVM technique uses six laser beams to illuminate droplets, which can then be easily imaged using the built-in CCD camera of the probe.

this technique [32]. It should be noted that the invasive nature of direct imaging techniques is significantly reduced by this method because of the 25 mm diameter of the probe.

4.2.4
Other Techniques

In addition to the approaches described earlier, there are other techniques that cannot be categorized into any group. These techniques have often introduced novel approaches or applied new methods, which require further attention in order to be employed in emulsification processes.

Bae and Tavlarides [11] employed a capillary method to determine the droplet size distribution of the mixing tank, which is illustrated in Figure 4.5. The system consists of a capillary tube positioned in the tank. Droplets during the process are forced into the capillary tube using a vacuum pump. Consequently, droplets form cylindrical slugs when they are drawn into the capillary tube. A laser beam is passed exactly across the middle of the tube. The device estimates the droplet size by measuring the difference between laser transmission caused by the difference in absorption of the solute and the droplets. One of the advantages of this technique is that it can be employed for high dispersed phase volume fractions. However, it is limited to materials that have low molar absorptivity. In addition, a calibration measurement is required to determine the relationship between the absorption and droplet size. This technique was modified by Hocq et al., [12] who, instead of using

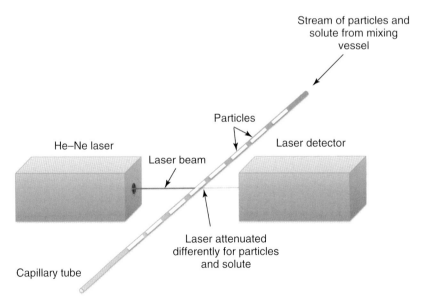

Figure 4.5 A capillary tube pulls particles from the mixing vessel, the laser is attenuated differently by the particles and the solvent, creating a signal in the detector, which can be used to extrapolate the droplet volume.

the difference in laser absorption, used the difference between the conductivity of the phases to determine the droplet sizes. Although the limitations of the technique with regard to the choice of materials are reduced by this modification, the other disadvantages of this technique remain.

Another approach was suggested by Hong and Lee [14], who employed a light transmission method to determine droplet size distribution. A fiber optic light guide received monochromatic light propagated from one end and transmitted the emergent light to the photocell. The probe had a 15 mm gap through which the emulsion sample passed. The relationship between the mean droplet size and the light transmission was determined using calibration studies on emulsions with known droplet sizes. Although this technique does not report the droplet size distribution, it can generate mean droplet size evolution data in the early stages of the process because of its fast data acquisition rate. The main drawback of this method is that the emulsions need to be transparent. Therefore, the technique cannot be used for emulsions with high dispersed phase volume fractions.

Hollingsworth *et al.* [13] developed a technique based on the nuclear magnetic resonance (NMR) pulse sequence Difftrain, which is able to measure the droplet size in the emulsification processes with dispersed phase volume fractions up to 20% during the mixing process. The mixing vessel was placed into an NMR spectrometer, which operated by pulsed-field gradient (PFG) NMR, while the NMR spectrometer was modified to determine the droplet size during the process. Normally, the data acquisition time for PFG NMR is approximately 20 min. However, by taking advantage of Difftrain, the acquisition time was markedly reduced to 3–10 s, rendering the method suitable for emulsification studies. This technique benefits from the fact that it is noninvasive and can operate on opaque systems. However, its development is still in the early stages of validation as the mixing vessel and the emulsion volume used are significantly different from those commonly used in industrial applications.

In the School of Chemical Engineering, in the University of Birmingham, a technique based on the relationship between light reflected (Y) from the emulsion and its properties was developed to determine the droplet size evolution in real time. Calibration curves were produced between emulsions with known droplet sizes and reflectance from emulsion for each of the dispersed phase volume fractions and emulsifier types. The examples of such calibration curves can be seen in Niknafs *et al.* [15] for emulsions containing 5, 10, 20, and 50% of the dispersed phase in the presence of Tween 20, and in Niknafs *et al.* [33] for emulsions containing 10% of dispersed phase in the presence of silica particles. All of the calibration curves show a linear dependency between droplet size and the reflectance in a semilogarithmic diagram. It was shown that the droplet size of the systems in the absence of added emulsifier can be estimated using Tween 20 calibration curves. This methodology was employed for investigating emulsification in an agitated vessel. A representation of the reflectance output as obtained from the chroma meter device is given in Figure 4.6a, which, as expected, shows that Y values change with changes to processing conditions; in this case the emulsification without any emulsifier started with 1600 rpm impeller speed for 2 h, then taken down to 800 rpm for 2 h

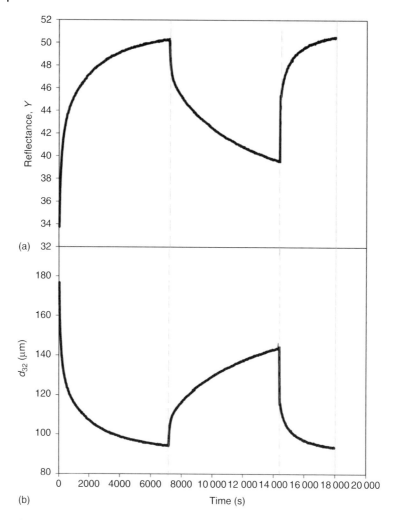

Figure 4.6 Representation of results obtained from the reflectance technique. (a) Reflectance values and (b) corresponding average droplet sizes (d_{32}) as a function of time. Dashed lines indicate changes in the processing conditions [15].

and finally increased to 1600 rpm for 1 h. The corresponding droplet size evolution can be then calculated using the obtained calibration curves; Figure 4.6b shows the droplet size data obtained from the reflectance measurements in Figure 4.6a.

The reflectance technique benefits from its online and noninvasive nature (it does not influence the process or formulation), fast data acquisition rate (which can be used for initial stages of the process, where the high rate of change in droplet size is expected), and simplicity (the technique can be used with relative ease). However, one of the main drawbacks of the technique relates to the fact that it does not provide any information regarding the droplet size distribution during processing and it requires calibration for each system that is to be studied.

4.3
Techniques Investigating Droplet Coalescence

Investigating droplet coalescence is more complex as it is more significant at higher dispersed phase volume fractions, thus limiting the use of online droplet size measurement techniques.

One of the first attempts was made by Howarth [34] who carefully induced a sudden change in the hydrodynamic condition of the process (a mixing vessel), which provided conditions under which droplet breakup would no longer affect the droplet size. Consequently, droplet coalescence dominated and increased the droplet size. This was achieved by a step-change reduction in the impeller speed. Consequently, the period directly following the impeller speed change was dominated by droplet coalescence. A similar method was employed by Wright and Ramkrishna [35], who determined the numerical values of droplet coalescence by systematic sampling and population balances. In addition, this approach was later employed by Mohan et al. [36] and Narsimhan and Goel [37] on homogenizer devices by inducing a sudden change in the hydrodynamic conditions of the process by a step-change reduction in the homogenization pressure. Although this technique can provide essential information, one of its main drawbacks is that droplet coalescence cannot be examined in the presence of droplet breakup [19].

This issue was resolved to a certain extent in an approach developed by Taisne et al. [18], who prepared two coarse emulsions (preemulsions) under similar conditions but with two different groups of oil phases (dispersed phase). One group was used in the natural state, while the other was brominated; thus, the two groups differed in their refractive index. In addition to these oils, sucrose was employed to match the refractive indices of the respective aqueous phases to those of the oil phases. Subsequently, these two emulsions were added together and the mixture was homogenized. By monitoring the refractive index of the homogenized emulsion, the extent of droplet coalescence was determined. If droplets coalesced, the refractive index of the homogenized emulsion would be bracketed between the refractive indices of the natural and brominated emulsions. Although this technique can indeed examine the droplet coalescence with droplet breakup, it suffers from a major problem. The refractive index of an emulsion depends on the droplet size; hence, the preemulsions were prepared by passing the emulsions through homogenizers until the droplet size was the same for both preemulsions. Therefore, the coalescence study was conducted on emulsions where the equilibrium droplet size had already been achieved. Another problem with this method is the chemical modification of the oil phases, which can indeed affect the physicochemical characteristics of the dispersed phase such as the interfacial tension.

A similar approach was followed by Lobo et al. [19]; however, instead of brominating the oil phase, they employed a fluorescent probe. This approach overcame the problem of dependency on the droplet size. Consequently, there was no requirement for keeping similar droplet sizes in the two preemulsions. Therefore, it was possible to examine droplet coalescence in the stages where a high frequency of droplet breakup is expected. In addition, a Monte Carlo simulation was employed to determine how random mixing affects the fluorescent signal. The simulation showed that the sensitivity of the technique can be tuned according to the process. Although this method resolved one of the major problems of the approach by Taisne *et al.* [18], it had a new limitation, that is, it could be employed only in emulsions that contain nonviscous dispersed phases.

Finally, a similar approach was proposed by Danner [20]. He proposed a coloring technique where preemulsions were prepared by employing dispersed phases with various hydrophobic colorants. Therefore, the change in the color of the droplets was linked with droplet coalescence. For example, red and blue dyes were employed to color the oil phases. Owing to droplet coalescence, the color of the droplets would change to purple. The color change was observed by using optical microscopy and, the coalescence was quantified by Monte Carlo simulation. An example of the image obtained from colored emulsions obtained in our laboratory can be seen in Figure 4.7.

Careful inspection of the obtained images (Figure 4.7) revealed the (over time) appearance of a third color (purple, droplet number 3 in Figure 4.7a) in the mixed systems. As the continuous phase of both the single- and the mixed-colored emulsions remains "colorless," the development of the third color via diffusion of one of the primary colors from one droplet (red or blue) through the continuous phase to another droplet (blue or red) can be dismissed. What can be therefore postulated is that the development of the observed third color can only be the result of two (or more) droplets of different initial colors, coalescing. Hence, this color changing process is directly related to the coalescence phenomena taking place in the system under observation.

The first drawback of coloring technique arises from the limitation of identification of the color of emulsion by optical observation. The color intensity is significantly reduced in the smaller droplets because of the presence of a lower amount of colorant and a higher curvature of droplets. This results in the merging of the color of smaller droplets in to the background (cream), which is the color

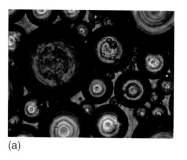
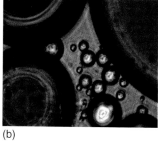

(a) (b)

Figure 4.7 Image captured from the samples obtained at the end of the mixed emulsion process. The emulsion contained 10% dispersed phase volume fraction in the presence of Tween 20 in an agitated vessel.

of the continuous phase. The experiments show that the smallest droplet that the color of which can be identified is 20 µm, but the droplet size distributions determined from a Mastersizer device (Malvern, UK) show that smaller droplets (10 µm) exist, which shows that a certain number of droplets are not included in the color quantification. Additionally, a major drawback of this technique is the amount of sampling involved. Although the last three above-mentioned approaches generate new insights into droplet coalescence, they require the preemulsions to be prepared.

4.4
Concluding Remarks

The evident remark that can be postulated from the presented review is that none of the above techniques, when used individually, are adequate for characterizing emulsions during processing. This indicates that combining different techniques might result in a more comprehensive experimental data on emulsification during processing. As an example, Hu *et al.* [38] employed the online imaging technique developed by Pacek *et al.* [10] for investigation of bubble coalescence by generating a coalescence dominant regime induced by a change in an impeller speed, which in turn is based on the technique developed by Howarth [17]. The combination of these two techniques results in better observation of droplet size change that in turn increases the accuracy regarding the characterization of the droplet coalescence. Niknafs *et al.* [15] and Niknafs *et al.* [33] employed a similar approach where the reflectance technique was used in combination with the methodology developed by Howarth [17]. An example of such an experiment can be seen in Figure 4.8, which shows the mean droplet size evolution of oil-in-water emulsions containing 10% of the dispersed phase in the presence of various emulsifier systems. The emulsification was conducted in a mixing tank.

It can be seen in Figure 4.8 that the change in impeller speed results in a droplet coalescence dominant regime. The detail discussion regarding the data can be seen in Niknafs *et al.* [33]. Nonetheless, it is evident that combining the two techniques can lead to a better understanding of emulsification.

In addition to the above-mentioned studies, several techniques can be employed for droplet size measurements. The selection of the techniques can be in such a way that the new technique utilizes the positive attributes of the selected techniques and overcomes the drawbacks of each technique when they are used individually. For example, we can consider combining the reflectance technique with the online imaging technique. The reflectance technique possesses some positive attributes that are required for online droplet size measurement, including fast data acquisition rate and ease of its implementation. However, it generates mean droplet size evolution data that cannot be used for a detailed study of droplet size distribution. In contrast, direct imaging provides droplet size distribution data, which could not be employed when time scales of the process are small. Therefore, a method based on the combination of both techniques can benefit from the fast

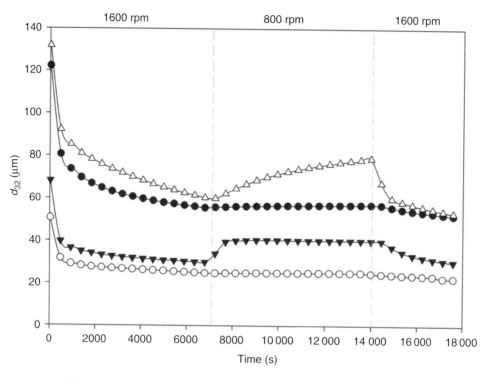

Figure 4.8 Mean droplet size evolution of oil-in-water emulsions containing 10% dispersed phase volume fraction in the presence of 1% Tween 20 (▼), 1% of silica particles (●), 1% of Tween 20, and 1% of silica particles (○) and in the absence of added emulsifier (Δ).

data acquisition rate of the reflectance technique and the droplet size distribution determination of the online imaging technique.

Most of the techniques measuring droplet size were employed for mixing systems and their applicability was not examined for other emulsification methods. These processes may include the homogenizer, membrane processes, and the newly developed impinging jet emulsification systems. The presented online droplet size measurement techniques can be used "inline" for homogenization process. For example, a number of studies were carried out by generating a process loop in which the outlet of a high pressure homogenizer was recycled back as a feed to the high pressure homogenizer by passing through a mixing vessel [18, 36, 37]. Subsequently, samples were obtained from the mixing tank in order to measure their droplet size, by using an offline particle size measurement device. This step can be avoided by implementing one of the online droplet size measurement techniques (e.g., the reflectance technique or PVM®) inline in the recycling loop. Currently, in the School of Chemical Engineering at the University of Birmingham we are trying to employ the developed reflectance technique for droplet size determination of the developing impinging jet emulsification process.

References

1. Henry, J.V.L., Fryer, P.J., Frith, W.J., and Norton, I.T. (2009) *J. Colloid Interface Sci.*, **338** (1), 201–206.
2. Alban, F.B., Sajjadi, S., and Yianneskis, M. (2004) *Chem. Eng. Res. Des.*, **82** (A8), 1054–1060.
3. Chatzi, E. and Lee, J.M. (1987) *Ind. Eng. Chem. Res.*, **26** (11), 2263–2267.
4. Chatzi, E.G., Boutris, C.J., and Kiparissides, C. (1991) *Ind. Eng. Chem. Res.*, **30** (3), 536–543.
5. Zhou, G.W. and Kresta, S.M. (1998) *Chem. Eng. Sci.*, **53** (11), 2099–2113.
6. Wille, M., Langer, G., and Werner, U. (2001) *Chem. Eng. Technol.*, **24** (5), 475–479.
7. Petrak, D. (2002) *Part. Part. Syst. Char.*, **19** (6), 391–400.
8. Alopaeus, V., Koskinen, J., Keskinen, K.I. and Majander, J. (2002) *Chem. Eng. Sci.*, **57** (10), 1815–1825.
9. Tong J. and Povey M.J.W. (2002) *Ultrasonics* **40** (1-8), 37–41.
10. Pacek, A.W., Moore, I.P.T., Nienow, A.W., and Calabrese, R.V. (1994) *AIChE J.*, **40** (12), 1940–1949.
11. Bae, J.H. and Tavlarides, L.L. (1989) *AIChE J.*, **35** (7), 1073–1084.
12. Hocq, S., Milot, J.F., Gourdon, C., and Casamatta, G. (1994) *Chem. Eng. Sci.*, **49** (4), 481–489.
13. Hollingsworth, K.G., Sederman, A.J., Buckley, C., Gladden, L.F., and Johns, M.L. (2004) *J. Colloid Interface Sci.*, **274** (1), 244–250.
14. Hong, P.O. and Lee, J.M. (1983) *Ind. Eng. Chem. Process Des. Dev.*, **22** (1), 130–135.
15. Niknafs, N., Spyropoulos, F., and Norton, I.T. (2011) *J. Food Eng.*, **104** (4), 603–611.
16. Vankova, N., Tcholakova, S., Denkov, N.D., Vulchev, V.D., and Danner, T. (2007) *J. Colloid Interface Sci.*, **313** (2), 612–629.
17. Howarth, W.J. (1964) *Chem. Eng. Sci.*, **19** (1), 33–38.
18. Taisne, L., Walstra, P., and Cabane, B. (1996) *J. Colloid Interface Sci.*, **184** (2), 378–390.
19. Lobo, L., Svereika, A., and Nair, M. (2002) *J. Colloid Interface Sci.*, **253** (2), 409–418.
20. Danner, T. (2001) Tropfenkoaleszenz in Emulsionen. PhD Thesis. Universitat Karlsruhe (TH).
21. Sis, H. and Chander, S. (2004) *Colloids Surf. A*, **235** (1–3), 113–120.
22. Maass, S., Grunig, J., and Kraume, M. (2009) *Chem. Process Eng.-Inz. Chem. I Procesowa*, **30** (4), 635–651.
23. Hu, B., Angeli, P., Matar, O.K., Lawrence, C.J., and Hewitt, G.F. (2006) *AIChE J.*, **52** (3), 931–939.
24. Worlitschek, J., Hocker, T., and Mazzotti, M. (2005) *Part. Part. Syst. Char.*, **22** (2), 81–98.
25. Tadayyon, A. and Rohani, S. (1998) *Part. Part. Syst. Char.*, **15** (3), 127–135.
26. Boscher, V., Helleboid, R., Lasuye, T., Stasik, B., and Riess, G. (2009) *Polym. Int.*, **58** (10), 1209–1216.
27. Dukhin, A. and Goetz, P. (2005) *Colloids Surf. A*, **253** (1-3), 51–64.
28. Chanamai, R. Coupland, J.N. and McClements, D.J. (1998) Colloids and Surfaces A: Physicochemical and Engineering Aspects **139** (2), 241–250.
29. Chanamai, R., Horn, G., and McClements, D.J. (2002) *J. Colloid Interface Sci.*, **247** (1), 167–176.
30. Richter, A., Voigt, T., and Ripperger, S. (2007) *J. Colloid Interface Sci.*, **315** (2), 482–492.
31. Galindo, E., Larralde-Corona, C.P., Brito, T., Córdova-Aguilar, M., Taboada, B., Vega-Alvarado, L., and Corkidi, G. (2005) *J. Biotechnol.*, **116** (3), 261–270.
32. O'Rourke, A.M. and MacLoughlin, P.F. (2005) *Chem. Eng. Process.*, **44** (8), 885–894.
33. Niknafs, N., Sppub_yearopoulos, F., and Norton, I.T. (2010). The dynamic behaviour of pickering emulsification. Proceedings of 5th World Congress on Emulsions, Lyon, France.
34. Howarth, W.J. (1967). *A.I.Ch.E. Journal*, **13** (5), 1007–1013.

35. Wright, H. and Ramkrishna, D. (1994) *AIChE J.*, **40** (5), 767–776.
36. Mohan, S. and Narsimhan, G. (1997) *J. Colloid Interface Sci.*, **192** (1), 1–15.
37. Narsimhan, G. and Goel, P. (2001) *J. Colloid Interface Sci.*, **238** (2), 420–432.
38. Hu, B., Nienow, A.W., and Pacek, A.W. (2003) *Colloids Surf. B*, **31** (1–4), 3–11.

5
Emulsification in Rotor–Stator Mixers
Andrzej W. Pacek, Steven Hall, Michael Cooke, and Adam J. Kowalski

5.1
Introduction

Rotor–stator mixers can be broadly characterized as energy-intensive mixing devices frequently used in the most demanding mixing processes. The main feature of these mixers is their ability to focus high energy/shear in a small volume of fluid. They are relatively inexpensive, versatile, energy efficient, and consist of a high-speed rotor enclosed in a stator, with the gap between them ranging from 100 to 3000 µm [1]. Typically, rotor tip speed is between 10 and 50 m s^{-1} [2], which, in combination with a very narrow gap, generates very high shear rates; therefore, these mixers are often called *high-shear mixers*. By operating at high speed, rotor–stator mixers can significantly reduce processing time and capital investment in manufacturing a range of different products. In terms of energy consumption per unit mass of product, rotor–stator mixers require high power input over a relatively short time; however, as the energy is uniformly delivered and dissipated in a relatively small volume, each element of fluid is exposed to a similar intensity of processing. Frequently, the quality of the final product is strongly affected by its structure/morphology and it is essential that the key ingredients are uniformly distributed throughout the whole mixer volume; for such cases, rotor–stator mixers are often more efficient than other types of mixing devices.

Rotor–stator devices are used for processing of both single-phase and multiphase systems. In single-phase systems, they are very efficient in blending of miscible liquids of very different viscosities. In multiphase systems, they are used to emulsify immiscible liquids [3], to deagglomerate and uniformly disperse nanoparticles in liquids [4, 5], and also to suspend fine air bubbles in dairy products [6]. Many everyday products that require high shear and high energy dissipation rates in their formulation are frequently manufactured using rotor–stator mixers, and typical examples of such processes and final products are briefly summarized in Table 5.1.

The most common application of rotor–stator mixers is in emulsification processes, where they are used to manufacture a wide range of emulsion-based

Emulsion Formation and Stability, First Edition. Edited by Tharwat F. Tadros.
© 2013 Wiley-VCH Verlag GmbH & Co. KGaA. Published 2013 by Wiley-VCH Verlag GmbH & Co. KGaA.

Table 5.1 Examples of products manufactured using rotor–stator mixers.

System	Process	Example
Single-phase liquids	Blending	Mixing very viscous fluids
Two-phase liquid–liquid	Emulsification	Products made of 1–20 μm drops Cosmetics: creams, deodorants Food: mayonnaise, sauces Pharma: drugs insoluble in water
Two-phase solid–liquid	Dissolution/dispersion	Formulation of required rheology Thickening of shampoos Suspensions of nanoparticles Toothpastes, paints
Two-phase gas–liquid	Dispersing air in viscous fluids	Foams, aerated food products Shower gels Ice creams

Source: Adapted from Ref. [7].

products with droplet sizes between 1 and 20 μm in food, pharmaceutical, and cosmetics industries to name the most common applications [8].

Other types of mixers used for emulsification such as valve homogenizers, ultrasound homogenizers, liquid whistles [3] are capable of manufacturing emulsions with similar droplet sizes. One of the main advantages of rotor–stator mixers, besides versatility and relatively low cost, is the range of scales available, which makes the same mixer type suitable for lab applications as well as for high-volume products.

The fact that rotor–stator mixers are so widely used in industry is somewhat surprising considering that the fundamental understanding of dispersion processes in such mixers is rather limited, and selection of the type of mixer is often based on tradition or on the manufacturer's advice, rather than on scientific description of the process. There is some very general information on rotor–stator mixers in open literature [2]. Until relatively recently, only a few papers on batch rotor–stator devices [9, 10] and on continuous rotor–stator mixers [11–14] have been published. The aim of this chapter is to help fill this gap and present a comprehensive review of currently available literature information on emulsification in rotor–stator mixers as well as very recent experimental and theoretical results obtained by the authors.

5.2
Classification and Applications of Rotor–Stator Mixers

There are a wide range of designs of rotor–stator mixers available from different manufacturers including Silverson [15], Ross [16], Chemineer [17], IKA Works

5.2 Classification and Applications of Rotor–Stator Mixers | 129

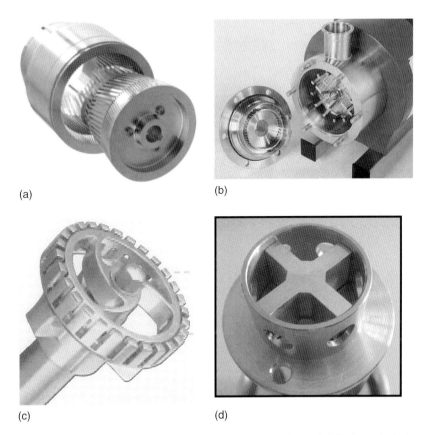

Figure 5.1 Rotor–stator mixer types: (a) colloid mill, (b) in-line radial discharge high-shear mixer, (c) toothed mixer, and (d) batch radial discharge mixer. (Source: Adapted from Refs. [7, 18, 21].)

[18], Rayneri [19], and Siefer [20] to name the largest ones. They can be broadly classified according to their mode of operation as batch or in-line (continuous) mixers. There are also other criteria that can be used for classification. According to the direction of the flow from the outlet of the mixer, they can be classified as radial and axial discharge mixers, and according to the design/shape of the rotor and stator, it is possible to distinguish between colloidal mills, toothed devices, and radial discharge high-shear mixers [2]. The most commonly used types of rotor–stator mixers are schematically shown in Figure 5.1.

5.2.1
Colloid Mills

Colloid mills are in-line mixers (Figure 5.1a) used for production of very fine liquid/liquid emulsions from premixed coarse emulsions often prepared in a stirred vessel, and also for deagglomeration and dispersion of different types of nanopowders, including pigments. They are characterized by a cone-shaped rotor and stator with smooth or grooved surfaces and a very narrow gap of variable width between them. The type of surface of the rotor and stator affects the flow regime with smooth surfaces promoting laminar flow and grooved surfaces promoting turbulent flow [1]. A variable gap width ensures velocity gradients in the direction of flow, which creates elongational stresses essential for breakage of very viscous drops. The average energy dissipation rate is rather high, of the order of $10^5 - 10^6$ W kg^{-1} [22], but pumping capability is very low; therefore, colloid mills typically operate as in-line mixers combined with external pumps [23]. Colloid mills have a conical rotor and stator, so the gap can be varied by moving the stator axially to close up or open up the gap so that the drop size is partly controlled by the rotor speed and partly by the width of the rotor–stator gap. Closing the gap leading to an increase of the shear rate was reported as one of the main advantages of colloid mills compared to other types of rotor–stator mixers [24], however, the reduction of the gap size does reduce throughputs. They are suitable for processing emulsions with a very high volume fraction of the dispersed phase such as bitumen emulsification [25], and have been used to manufacture mayonnaise containing 80% oil dispersed in water with egg yolk emulsifier [26]. Colloid mills were also employed for the manufacturing of water-in-oil emulsions, such as environmentally friendly fuel for heavy-duty internal combustion engines, reducing emissions of fine solid particles and NO$_x$ [27].

5.2.2
In-Line Radial Discharge Mixers

In-line radial discharge mixers (Figure 5.1b) are characterized by high throughput and good pumping capacity at low energy consumption. They are very flexible and frequently recommended as mixing devices in processing lines used to manufacture different products [24]. These mixers are relatively inexpensive and simple in operation compared to other emulsification devices such as high-pressure homogenizers, and are generally considered robust and reliable in factory environments (D. Rothman, private communication). In this type of mixer, the dispersed phase can be injected directly into the high-shear/turbulent zone where mixing is much faster than by injection into the pipe or into the holding tank. This might be important in systems with fast chemical reactions or when low-density solids are added to the liquid phase [24]. Typically in-line radial discharge mixers are supplied as single rotor–stator units but for more demanding processes such as dispersion of high-viscosity oils, multiple rotor–stator combinations are also available. Typically, multiple rotor–stators generate higher shear and more intense turbulence, therefore allowing the possibility to shorten processing time and enable more accurate control of droplet size. They are used for manufacturing of both emulsions and suspensions of very fine drops/solid particles of relatively narrow dispersed phase size distributions, as the design is such that all of the process streams pass through the region of shear/turbulence of similar intensity. They are typically supplied with a range of interchangeable screens, making them versatile and flexible in different applications, therefore finding use in a wide range of industries. In the petroleum industry, Silverson mixers have been used in pilot plant emulsification of hard and soft bitumen in aqueous continuous phases at pH 2 [28]. In the chemical industry, in-line mixers have been used to disperse highly concentrated insoluble ingredients in the continuous aqueous phase in the production of pesticides and herbicides, and also for dispersion of polymethacrylates and styrene-based copolymers to improve viscosity control of lubrication oils [15].

Hygienic versions of radial discharge rotor–stator mixers are widely used in personal care, food, and pharmaceutical industries. In the personal care product industry, they are used to disperse complex surfactants forming intermediary, difficult-to-disperse phases, and also to disperse waxes and pigments in lipstick manufacture. Topical pharmaceutical creams and heat-sensitive active ingredients of sterile ointments are also manufactured with the help of rotor–stator mixers, and dissolution of high-concentration sugars and thickeners during preparation of cough mixtures and pharmaceutical syrups are also carried out in those devices. In the food and beverage industry, they have been employed to disperse artificial sweeteners and clouding agents in fizzy drinks, as well as gums, starches, and alginates in fruit juices and carbonated drinks as thickeners [15], and also for processing of mayonnaise and salad dressings [29].

5.2.3
Toothed Devices

Toothed devices (Figure 5.1c) are available as in-line as well as batch mixers, and because of their open structure they have a relatively good pumping capacity; therefore, in batch applications, they frequently do not need an additional impeller to induce bulk flow even in relatively large mixing vessels. The majority of manufacturers offer a wide range of rotor–stator designs tailored to applications in emulsification in liquid–liquid systems and in a range of processes in solid–liquid systems such as incorporation, wet milling, and homogenization of solid particles in liquids. These mixers are used in the food industry to manufacture ice cream, margarine, and salad dressings; in cosmetic and personal care products to manufacture creams and lotions; and in manufacturing of speciality chemicals for microencapsulation of waxes and paraffin. They are also popular in the paper industry to process highly viscous and non-Newtonian paper pulp and in the manufacturing of paints and coatings [18]. It has been reported that Ultra-Turrax mixers have been used to manufacture emulsion-based lipid carriers with drops of size below 1 μm [30] and in emulsion polymerization to produce drops of the order of 300 nm [31]. Emulsions with such small drops are typically manufactured using less energy efficient ultrasound devices and these results indicate that in many processes, especially in pharmaceutical emulsions where drop size cannot exceed 5 μm, ultrasound emulsifying devices can be replaced by cheaper and more efficient rotor–stator mixers. Maa and Hsu [32] used a flow-through toothed mixer to emulsify an oil phase in aqueous bovine serum albumin to manufacture primary emulsions in the double-emulsion microencapsulation process. Masmoudi et al. [33] showed that this type of rotor–stator mixer can be used to prepare five basic pharmaceutical and cosmetic oil-in-water emulsions, representing pharmaceutical and cosmetic creams, whereas Yuan et al. [34] used a rotor–stator homogenizer to prepare a model fragrance encapsulated in polymerized monomers. In petroleum processing, toothed mixers were used to emulsify an aqueous phase containing nonionic surfactants in standard low-sulfur diesel oil, to form a water-in-diesel emulsion identical to a commercial European diesel formulation, characterized by drastic reductions of black smoke and NO_x emissions in car exhausts [35].

5.2.4
Batch Radial Discharge Mixers

Batch radial discharge mixers such as Silverson and Ross mixers (Figure 5.1d) have a relatively simple design with a rotor equipped with four blades pumping the fluid through a stationary stator perforated with differently shaped/sized holes or slots. They are frequently supplied with a set of easily interchangeable stators enabling the same machine to be used for a range of operations such as emulsification, homogenization, blending, particle size reduction, and deagglomeration. Changing from one screen to another is quick and simple. Different stators/screens used in batch Silverson mixers are shown in Figure 5.2. The general-purpose disintegrating stator (Figure 5.2a) is recommended for disintegration/solution of solids, preparation of gels and thickeners, and processing slurries while the slotted disintegrating stator (Figure 5.2b) is designed for disintegration of fibrous materials such as animal and vegetable tissue, and elastic materials such as rubbers and polymers. Square-hole screens (Figure 5.2c) are recommended for the rapid size reduction of soluble and insoluble solid powders and are also suitable for the preparation of emulsions and fine colloidal suspensions, whereas the standard emulsor screen (Figure 5.2d) is used for liquid–liquid emulsification.

Radial discharge, high-shear mixers are used in a wide range of industries ranging from foods through to chemicals, cosmetics, and pharmaceutical industries. In food processing, they can be used for manufacturing sweets as well as pet foods and sauces. In advanced food processing, microgel beads used as encapsulation vehicles for a wide range of hydrophilic nutraceuticals and fat replacers were manufactured

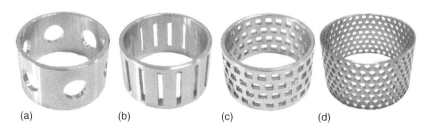

Figure 5.2 (a–d) Stators used in batch Silverson radial discharge mixers [15].

by hot emulsification of an aqueous solution of j-carrageenan in rapeseed oil using a Silverson mixer [36]; protein-stabilized oil-in-water emulsions used in various food applications can also be prepared in the same type of mixers [37]. Batch Silverson rotor–stator mixers were also used to deagglomerate and uniformly disperse dye nanoparticles and silica nanoparticles in aqueous phases containing different types of surfactants [4, 5]. In similar applications, Silverson rotor–stator mixers are used in cosmetic and pharmaceutical industries to manufacture both concentrated liquid–liquid and liquid–solid emulsions such as creams, lotions, mascaras, and deodorants to name the most common applications.

5.2.5
Design and Arrangement

Batch-toothed and radial discharge rotor–stator mixers are manufactured in different sizes ranging from the laboratory to the industrial scale. In lab applications, mixing heads (assembly of rotor and stator) can be as small as 0.01 m (Turrax, Silverson) and the volume of processed fluid can vary from several milliliters to a few liters (see Figure 5.1b,d). In models used in industrial applications shown in Figure 5.3, mixing heads might have up to 0.5 m diameter, enabling processing of several cubic meters of fluids in one batch.

As mentioned above, the understanding of mixing/dispersive processes in rotor–stator mixers is still rather limited; therefore, the geometry of Silverson rotors and stators at large scale are geometrically identical to the rotor and stators used at lab scale shown in Figure 5.2. Also the shape of Turrax rotor–stator assemblies is practically identical at different scales.

In practical applications, the selection of the rotor–stator mixer for a specific emulsification process depends on the required morphology of the product, frequently quantified in terms of average drop size or in terms of drop size distributions (DSDs) and by the scale of the process. As discussed in Section 5.3, in open literature, there is very little information enabling calculation of average drop size in rotor–stator mixers and there are no methods enabling estimation of DSDs. Therefore, the selection of an appropriate mixer and processing conditions for a required formulation is frequently carried out by trial and error. Many leading manufacturers offer their mixers either for a trial period or they can carry out lab-scale emulsification of given formulations [15] testing different type/geometries of mixers they manufacture. Obviously, the latter option might not be acceptable to many companies because of the commercial sensitivity of the processes they use for manufacturing emulsion-based products. Once the type of mixer and its operating parameters are determined at the lab scale, the process needs to be scaled up and recent research into scaling up of rotor–stator mixers carried out by the authors is discussed in the following sections.

The majority of lab tests of emulsification is carried out in small batch vessels as it is easier and cheaper than running continuous processes; therefore, prior to scaling up of the rotor–stator mixer, it has to be decided whether industrial emulsification should be run as a batch or as a continuous process. Batch mixers are recommended

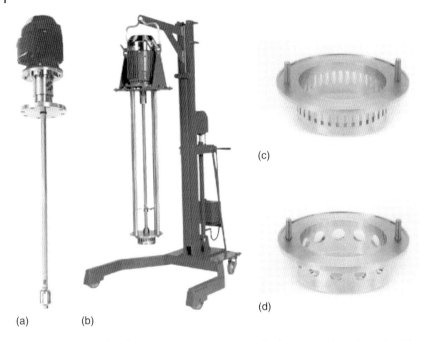

Figure 5.3 Industrial-scale rotor–stator mixers: (a) toothed mixer – IKA medium-size Ultra Turrex for up to 6 m^3 batches; (b) large-size batch radial discharge Silverson mixer, batches up to 35 m^3, (c) slotted screens (up to 0.5 m diameter) used in mixer (b), and (d) disintegrating screens used in mixer (b).

for processes where formulation of a product requires long processing times typically associated with slow chemical reactions. They require simple control systems, but spatial homogeneity may be an issue in large vessels, which could lead to a longer processing time. In processes where quality of the product is controlled by mechanical/hydrodynamic interactions between continuous and dispersed phases or by fast chemical reactions, but large amounts of energy is necessary to ensure adequate mixing, in-line rotor–stator mixers are recommended. In-line mixers are also recommended to efficiently process large volumes of fluid.

In the case of batch processing, rotor–stator devices immersed as top-entry mixers (Figure 5.4a) is mechanically the simplest arrangement but in some processes bottom-entry mixers (Figure 5.4b) ensures better bulk mixing, however, in this case, sealing is more complex. In general, the efficiency of batch rotor–stator mixers decreases as the vessel size increases and as the viscosity of the processed fluid increases because of limited bulk mixing by the mixers. While the open structure of Turrax mixers (Figure 5.1c) frequently enables sufficient bulk mixing even in relatively large vessels, if the liquid/emulsion has a low apparent viscosity, processing of very viscous emulsions requires an additional impeller (typically anchor type) to induce bulk flow and to circulate the emulsion through the rotor–stator mixer. On the other hand, batch Silverson rotor–stator mixers have a

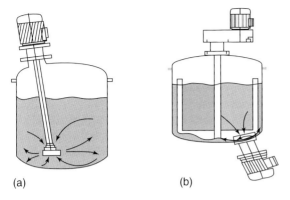

Figure 5.4 (a,b) Possible arrangements of rotor–stator mixers for batch processing.

very limited pumping capacity, and even at the lab scale, they are mounted off the center of the vessel to improve bulk mixing. At the large scale, there is always need for at least one additional impeller (Figure 5.3b), and in the case of very large units, more than one impeller is mounted on the same shaft.

There are also mechanical problems associated with fitting large, heavy mixers in tall vessels. Firstly, the weight of the motor/shaft assembly requires substantial support and secondly, long shafts rotating at thousands of revolutions per minute might need extra support/bearings to avoid oscillations and wobbling. These problems might be avoided by mounting the mixer at the bottom of the vessel (Figure 5.4b) with a rather short, more stable shaft, but in such cases appropriate seals are necessary. Also, close proximity of the mixer from the bottom of the vessel further limits its pumping capacity; therefore, in such configurations, additional impellers are necessary for bulk mixing.

Problems associated with the application of batch rotor–stator mixers for processing large volumes of fluid discussed above can be avoided by replacing batch mixers with in-line (continuous) mixers briefly discussed above. There are many designs offered by different suppliers (Silverson, IKA, etc.), and the main differences are related to the geometry of the rotors and stators with stators and rotors designed for different applications. The main difference between batch and in-line rotor–stator mixers is that the latter have a strong pumping capacity, and therefore they are mounted directly in the pipeline. One of the main advantages of in-line over batch mixers is that for the same power duty, a much smaller mixer is required, and therefore they are better suited for processing of large volumes of fluid [38]. When the scale of the processing vessel increases, a point is reached where it is more efficient to use an in-line rotor–stator mixer rather than a batch mixer of a large diameter. Because power consumption increases sharply with rotor diameter (to the fifth power), an excessively large motor is necessary at large scales. The transition point depends on the fluid rheology, but for a fluid with a viscosity similar to water, it is recommended to change from a batch to an in-line rotor–stator process at a volume of approximately 1–1.5 t (D. Rothman, private communication).

Figure 5.5 In-line Silverson rotor–stator geometry (64 mm diameter): (a) single-stage rotor and (b) single-stage stator; (c) double-stage rotor and (d) double-stage stator.

The majority of manufacturers supply both single and multistage mixers with the latter being recommended for more demanding mixing processes such as emulsification of highly viscous liquids or processing/size reduction in solid/liquid systems. Examples of single-stage and two-stage rotor–stator mixers are shown in Figure 5.5.

5.2.6
Operation

In-line mixers can be used in a single-pass mode for continuous blending of several miscible liquids of very different viscosities (Figure 5.6a) or for systems where fast chemical reaction occurs. In this configuration, they are also common in manufacturing of liquid detergents as aeration, which can adversely affect the quality of products containing substantial amounts of surfactant, is minimal. Typically industry is moving toward single-pass systems as they are more attractive in terms of reducing energy consumption and manufacturing timescales. In general, multistage mixers as opposed to single-stage mixers are recommended

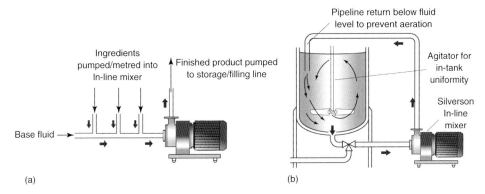

Figure 5.6 In-line rotor–stator mixers: (a) single-pass and (b) multipass arrangements [39].

for single-pass operations to enhance droplet breakage during the relatively short residence time available (D. Rothman, private communication).

In more demanding processes such as those with high degree of size reduction of drops or solid aggregates, multipass systems (recirculation) are frequently used (Figure 5.6b). In such systems, the morphology of the product (drop/particle size distributions) is not only controlled by the rotor speed and total flow rate but also by the number of passes (residence time) through rotor–stator mixer. This allows for further droplet breakage or fluid structuring. Multipass processing has proved to be particularly attractive in deagglomeration and uniform dispersion of nanopowders in liquids. At very large scale in such configurations, the bulk mixing in stirred vessels is intensified by additional impellers if necessary.

In multipass applications of rotor–stator mixers, a holding tank and a closed loop pump are required to circulate the processed materials. This increases capital costs and requires more sophisticated control systems since both flow rate and rotor speed need to be controlled to achieve the required degree of emulsification, or in case of solid aggregate size reduction [40]. However, in many industrial processes, it is believed that multipass arrangements should be used to achieve a product specification but very often it is more a precaution than real need; and again, such unnecessary processing can only be explained by a lack of understanding of mixing/dispersion processes in rotor–stator mixers.

The above review of rotor–stator mixers is far from complete because a large number of different designs are available from major manufacturers. Also, because of their simplicity and the number of available design changes – and frequently only minor changes in design – lead to "new, better, more efficient … " mixers coming to the market. The main aim of this review was to give the reader a broad picture of the types of rotor–stator mixers that are currently available on the market and to point out the main areas of processing of complex multiphase systems where rotor–stator mixers are successfully employed.

5.3
Engineering Description of Emulsification/Dispersion Processes

It has been frequently postulated that high shear is the main driving force for very efficient single-phase mixing (blending) as well as multiphase dispersion (emulsification, deagglomeration) in rotor–stator mixers; therefore, rotor–stator mixers are frequently called *high-shear mixers*. This hypothesis was questioned relatively recently: first, indirectly, when the results of experimental investigations indicated that drops are broken by turbulent velocity and pressure fluctuations rather than shear stress [11, 41]; second, by direct confirmation of the strong effect of turbulence on mixing and dispersion processes from numerical simulations carried out using computational fluid dynamics (CFD) of single-phase flow inside rotor–stator mixers and verified by measurements of velocity distributions in close proximity to the stator [9, 10] and from numerical simulation of two-phase liquid–solid flow [14]. It is now commonly accepted that high-intensity turbulence generated inside the stator leading to very high local energy dissipation rates inside and in very close proximity to the stator strongly affects both mixing and dispersion processes.

Efficiency and kinetics of dispersion processes in two-phase systems such as emulsification in liquid–liquid systems or breakage of agglomerates (deagglomeration) in solid–liquid systems are determined by the interaction between the dispersed and continuous phases, which in turn is determined by the type and intensity of flow. There are well-established, mechanistic models relating drop size and mechanisms of their breakage to the flow regime and physical/interfacial properties of both phases. Despite the fact that flow in rotor–stator mixers is very complex, and the models used to describe drop breakage are developed for ideal laminar and turbulent flows, experimental results indicate that those simple models frequently describe average drop size produced in rotor–stator mixers rather well; therefore, they are briefly discussed below.

5.3.1
Drop Size Distributions and Average Drop Sizes

The great majority of emulsifying devices produce polydispersed emulsions with drops within a certain size range; therefore, accurate description of the size of the dispersed phase requires knowledge of the DSD. Detailed discussion of the concept of DSDs is well outside the scope of this chapter and it is enough to say that practically all instruments used to measure drop size show both differential and cumulative DSDs. An example of such distributions measured by a Malvern Mastersizer particle analyzer showing the volume of dispersed phase as a function of drop size is shown in Figure 5.7.

The functions shown above enable the detailed description of the population of drops/particles such as volume, area, and number of drops of different sizes. In some cases, knowledge of DSDs is essential, for example, in pharmaceutical emulsions, drop size cannot exceed approximately 5 μm, and frequently the size of

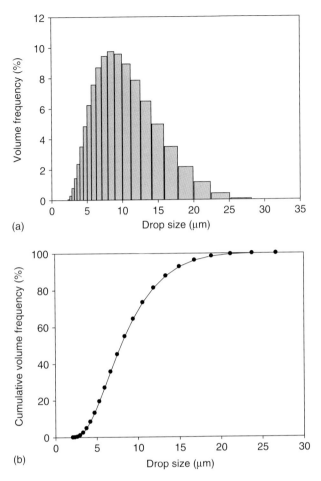

Figure 5.7 Drop size distributions: (a) volume frequency in discrete classes and (b) cumulative volume distribution.

the smallest drop is also limited to tens of nanometers. However, in many practical applications, rather than using DSDs (functions) to describe the population of drops, a certain mean size (number) is frequently used. *Mean drop size* is defined as the ratio of selected moments of the size distribution:

$$d_{pq} = \left[\frac{\int_0^\infty d^p f(d) \partial d}{\int_0^\infty d^q f(d) \partial d} \right]^{1/(p-q)} \tag{5.1}$$

where p and q are integers and $p > q$ and typically p does not exceed 4.

Equation (5.1) shows that it is possible to define several mean/average sizes for the same population of drops having completely different values especially if the DSD is wide. One of the means commonly used in emulsification/dispersion processes is Sauter mean diameter:

$$d_{32} = \frac{\int_0^\infty d^3 f(d) \partial d}{\int_0^\infty d^2 f(d) \partial d} \quad (5.2)$$

that relates specific interfacial area to the volume of the dispersed phase:

$$a = 6 \frac{\phi}{d_{32}} \quad (5.3)$$

The other two commonly used mean diameters are number mean:

$$d_{10} = \left[\frac{\int_0^\infty d^1 f(d) \partial d}{\int_0^\infty f(d) \partial d} \right] \quad (5.4)$$

and mass mean:

$$d_{43} = \left[\frac{\int_0^\infty d^4 f(d) \partial d}{\int_0^\infty d^3 f(d) \partial d} \right] \quad (5.5)$$

In this chapter, unless stated otherwise, the Sauter mean diameter d_{32} is used.

5.3.2
Drop Size in Liquid–Liquid Two-Phase Systems – Theory

Despite extensive research on dispersed liquid–liquid systems, there are no models enabling theoretical prediction of DSDs, instead simple mechanistic models allowing estimation of maximum stable drop size were developed. The maximum stable drop size that can exist in certain hydrodynamic conditions can be estimated considering the cohesive (τ_c) and disruptive (τ_d) stresses acting on the drop. In general, cohesive stress depends on interfacial tension (σ), viscosity of the dispersed phase (μ_d), and radius of the drop (r); for nonviscous dispersed phases, it is frequently approximated by

$$\tau_c \approx \frac{\sigma}{r} \quad (5.6)$$

Disruptive stress depends on viscosity of the continuous phase (μ_c), radius of the drop (r), type of flow (laminar or turbulent), and its intensity described either by shear rate ($\dot{\gamma}$) in laminar flow or specific energy dissipation rate (ε) in turbulent flow. The drop is stable if $\tau_d < \tau_c$ and maximum stable drop size can be calculated from:

$$\tau_d = \tau_c \quad (5.7)$$

The ratio of disruptive to cohesive stress τ_d/τ_c is called the *capillary number* (*Ca*) if the flow is laminar, and *Weber number* (*We*) if the flow is turbulent, and it is commonly used to relate drop size to hydrodynamic conditions and physical properties of the dispersed and continuous phases. The majority of correlations for maximum stable drop size published in literature are based on the above simple mechanistic model.

5.3.3
Maximum Stable Drop Size in Laminar Flow

To the first approximation, *laminar flow* can be defined as flow where fluid moves in separated layers and where there is no flow in the direction perpendicular to those layers. The following types of such flows might occur in processing equipment: simple shear flow, elongational flow, and hyperbolic flow.

Simple shear flow occurs in the fluid contained between two concentric cylinders separated by a narrow gap with both rotating in opposite directions or with one rotating and one stationary. This type of flow might approximate flow in the gap of radial discharge rotor–stator mixers. Elongational flow occurs when there is a velocity gradient in the direction of flow and it might occur in colloid mills when the cross section of the gap varies. Hyperbolic flow can be used to approximate a jet of liquid hitting the surface, and to the first approximation, it can be used to model flow of very viscous emulsions in high-pressure homogenizers. In all types of flow, drops are deformed and eventually broken by shear stress resulting from velocity gradients across the drop:

$$\tau_d = \mu_c \frac{dU_i}{dx_j} = \mu_c \dot{\gamma} \qquad (5.8)$$

The capillary number at which breakage occurs defines maximum stable drop size and it is called the *critical capillary number*:

$$Ca_{cr} = \frac{\dot{\gamma} \mu_c d_{max}}{\sigma} \qquad (5.9)$$

The critical capillary number in laminar flow depends on the type of flow and viscosity ratio $\lambda = \mu_d/\mu_c$ and it can be determined from stability curves (also called *Grace curves*) shown in Figure 5.8.

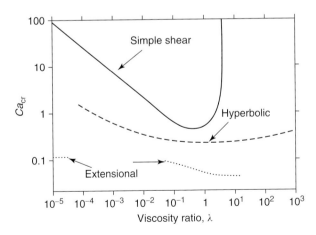

Figure 5.8 Critical capillary number for different types of laminar flow (Source: Adapted from Ref. [42].)

If the viscosities of both phases, interfacial tension, and shear rate are known, the above curves can be used to find the maximum stable drop size:

$$d_{max} = Ca_{cr}\sigma/\dot{\gamma}\mu_c \qquad (5.10)$$

and the Sauter mean diameter:

$$d_{32} \approx d_{max} \approx \dot{\gamma}^{-1} = C_1 \dot{\gamma}^{-1} \qquad (5.11)$$

From Figure 5.8, it is obvious that elongational and hyperbolic flows are more efficient for breakage then simple shear flow, and very viscous drops (viscosity ratio $\lambda > 4$) cannot be broken in simple shear flow even at very high shear rates. Therefore, such types of flow are unsuitable for processing of emulsions with very viscous dispersed phases. Equation (5.11) implies that if the size of drops produced by an emulsifying device is inversely proportional to the shear rate, it is very likely that the drops are broken by shear stress.

5.3.4
Maximum Stable Drop Size in Turbulent Flow

Turbulent flows can be characterized by chaotic, random changes of fluid velocity both in space and time. Analysis of velocity distributions in such flows is very complex; therefore, turbulent flows in emulsification devices are commonly described in terms of energy dissipation (scalar) rather than in terms of local velocity (vector) by the well-known Kolmogorov model. This model assumes that homogeneous, isotropic turbulence can be described by an energy cascade that is modeled as a population of differently "sized" eddies ("size" of an eddy is a measure of energy it contains) undergoing continuous breakage associated with energy transfer from the largest (energy-containing) eddies that break into smaller ones without losing energy. The smallest eddies corresponding to Kolmogorov's length scale cannot be broken any further, and at that scale, mechanical energy is dissipated into heat through viscous dissipation. During emulsification, since the amount of energy transferred from the dispersed to the continuous phase determines the maximum stable drop size, this model is well suited to describe dispersion processes [43].

Within this model, stress acting on drops depends on the size of the drop relative to the "size" of eddies and to the Kolmogorov length scale:

1) Drops larger than Kolmogorov's length scale are "bombarded by eddies" of size similar to the drop size, which results in normal disruptive stresses acting on the drop.
2) Drops smaller than Kolmogorov's length scale are immersed in the smallest eddies and they are exposed to disruptive shear stresses resulting from flow within the eddies.

On the basis of these assumptions, Hinze [43] developed expressions relating maximum stable drop size to local energy dissipation. Drops larger than Kolmogorov's length scale are exposed to disruptive normal stresses that can be related to the local energy dissipation rate:

$$\tau_d \approx \rho_c(\varepsilon d)^{2/3} \tag{5.12}$$

Therefore, maximum stable drop size and Sauter mean diameter can also be related to average energy dissipation rate:

$$d_{max} = C_2 \sigma^{0.6} \varepsilon^{-0.4} \rho_c^{-0.6} \tag{5.13}$$

$$d_{32} = C_3 \varepsilon^{-0.4} \tag{5.14}$$

Drops smaller than the Kolmogorov's length scale are exposed to the shear stress inside the smallest eddies that can also be related to local energy dissipation rate [44]:

$$\dot{\gamma} = C_4 \left(\frac{\varepsilon}{\nu_c}\right)^{0.5} \tag{5.15}$$

and in such a case,

$$d_{32} = C_5 \dot{\gamma}^{-1} = C_6 \varepsilon^{-0.5} \tag{5.16}$$

In summary, the average drop size in laminar flow is related to shear rate that can be estimated if the local shear rate (velocity gradient $\dot{\gamma}$) and geometry of the flow are known, whereas average drop size in turbulent flow can be estimated if the local energy dissipation rate (ε) is known.

5.3.5
Characterization of Flow in Rotor–Stator Mixers

In rotor–stator mixers, both shear rate in laminar flow and energy dissipation in turbulent flow strongly depend on the position inside the mixer. Until recently, it was impossible to predict their spatial distributions; therefore, an average shear rate or average energy dissipation rate in the swept volume of the impeller or the rotor has been commonly used to quantify intensity of flow. This approach, which is based on the description of mixing/dispersion processes in stirred vessels, was developed over the past 50 years, and is discussed below, whereas more advanced methods based on CFD developed recently for rotor–stator mixers are briefly discussed in Section 5.4.

5.3.5.1 Shear Stress
In laminar flow in stirred vessels, the average shear rate is proportional to the rotor speed with the proportionality constant K dependent on the type of impeller [45]:

$$\dot{\gamma} = KN \tag{5.17}$$

In stirred vessels, the proportionality constant cannot be calculated and has to be determined experimentally. In rotor–stator mixers, the average shear rate in the

Table 5.2 Distribution of energy dissipated in and around a radial discharge batch Silverson mixer.

Type of screen	Rotor-swept volume (%)	Hole region (%)	Jet region (%)	Rest of the vessel (%)
Disintegrating (Figure 5.2b)	48.5	7.8	21.0	22.8
Slotted (Figure 5.2a)	56.3	13.4	24.2	6.3
Square-hole (Figure 5.2c)	60.5	11.5	26.6	1.4

Source: Adapted from Ref. [7].

gap between the rotor and stator can be calculated if the rotor speed and geometry of the mixer are known:

$$\dot{\gamma} = \frac{\pi DN}{\delta} = K_1 N \tag{5.18}$$

Theoretically, in this case, there is no need for experimental data. However, it has been reported that K_1 depends on fluid rheological properties; therefore, it should also be determined experimentally [46].

5.3.5.2 Average Energy Dissipation Rate

The average energy dissipation rate in turbulent flow in rotor–stator mixers can be calculated from

$$\varepsilon = \frac{P}{\rho^c V} \tag{5.19}$$

if the power draw (P) and the volume the energy is dissipated in are known. Selection of the volume in which the energy is dissipated is not obvious as the spatial distribution of energy dissipation rate is strongly affected by the type of stator. Table 5.2 shows calculated distributions of energy dissipation rates around a small-scale, radial discharge batch Silverson mixer (Figure 5.1d).

From the above table, it is clear that the definition of volume for calculation of the average energy dissipation rate is far from obvious and, typically, it is assumed that all energy supplied by the rotor is dissipated within a certain arbitrary volume around the mixing head or in the swept volume of the rotor.

5.3.5.3 Power Draw

The power draw in batch rotor–stator mixers is calculated in the same way as in stirred vessels [47]:

$$P = Po \rho_c N^3 D^5 \tag{5.20}$$

While in the case of stirred vessels, the power number characterizes the impeller [45], in the case of rotor–stator mixers it appears to be a feature of the rotor–stator combination. Padron [47] reported *Po* for a radial-flow Ross mixer between 2.4 and 3.0 and for a Silverson mixer between 1.7 and 2.3. He also found that in turbulent

5.3 Engineering Description of Emulsification/Dispersion Processes

flow, the effect of the rotor–stator gap width on the power number is marginal and that the energy is mainly dissipated in turbulent jets emerging from the stator openings. This finding was later confirmed by CFD simulations and experimentally measured velocity distributions using laser Doppler anemometry (LDA) in close proximity to the stator of the Silverson mixer [9, 10].

In in-line rotor–stator mixers, power draw has been described in a similar way as power draw in centrifugal pumps and likened with the kinetic energy of the fluid [48]:

$$P = Po \rho_c Q N^2 D^2 \tag{5.21}$$

This equation indicates that at zero flow rate the calculated power draw is zero, which is clearly incorrect.

A more accurate expression for power draw by in-line rotor–stator mixers in turbulent flow was developed by Baldyga et al. [13] and Kowalski [38], and validated by Kowalski et al. [49]:

$$P = Po_z \rho_c N^3 D^5 + k_1 M N^2 D^2 + P_L \tag{5.22}$$

The first term is analogous to power consumption in a batch rotor–stator mixer, and the second term takes into account the effect of the pumping action on total power consumption [38]. The third term accounts for mechanical losses and is typically below a few percent, and therefore can be ignored. For in-line Silverson rotor–stator mixers fitted with double rotors and standard double-emulsor stators, power draw predicted from the constants, $Po_z = 0.197$ and $k_1 = 9.35$ calculated by torque, are in very good agreement with measured energy dissipation rates up to 80 000 W kg^{-1} [49].

While in turbulent flow, Po_z in Eq. (5.22) is approximately independent of the Re number, in laminar flow there is a strong dependency of power number on Re, and in this case, power draw can be calculated from [50]

$$P = k_0 N^2 D^3 \mu^c + k_1 M N^2 D^2 + P_L \tag{5.23}$$

where k_0 is a constant and depends on the Re number:

$$Po_z Re = k_0 \tag{5.24}$$

From Eqs. (5.19)–(5.24) the average energy dissipation rate in the rotor–stator can be calculated and used in semiempirical correlations for average drop size discussed below.

5.3.6
Average Drop Size in Liquid–Liquid Systems

Theoretical models relating average drop size to an average energy dissipation rate (Eq. (5.13)) can be rearranged to the dimensionless form

$$\frac{d_{32}}{D} = C_7 We^{-0.6} \tag{5.25}$$

This equation is commonly used to correlate drop size in dilute liquid–liquid dispersions processed in stirred vessels where the proportionality constant C_7 is in the range 0.09–0.125 [51]. It is determined experimentally and depends on the type of impeller. The accuracy of this type of correlation can be improved by accounting for the effect of intermittency on the disruptive stress acting on drops [52].

Calabrese et al. [41] employed the same model to describe average drop size in a batch rotor–stator mixer. They investigated the effect of the stator geometry, volume fraction, and interfacial tension of nonviscous dispersed phases on the average drop size in a Ross mixer and found that the Sauter mean diameter is well correlated by the following equation:

$$\frac{d_{32}}{D} = 0.04 \, We^{-0.58} \tag{5.26}$$

For the same type of mixer but for a wider range of viscosities of the dispersed phase (0.5–50 mPa·s) and interfacial tension (4.5–51 mN m^{-1}), Puel et al. [53] found that the average drop size also depends on viscosity ratio:

$$\frac{d_{32}}{D} = 0.02 \, We^{-0.6} \left(\frac{\mu_d}{\mu_c}\right)^{0.5} \tag{5.27}$$

The average drop size in a pilot plant scale, in-line Silverson rotor–stator mixer was also correlated by a similar equation. Hall et al. [11] investigated the effect of flow rate (up to 4800 kg h^{-1}) and viscosity (9–970 mPa·s) on the average drop size of silicone oil in water containing surface-active agent preventing coalescence, and found that the theoretical value of the exponent on energy dissipation rate gives a very good fit to the experimental data:

$$\frac{d_{32}}{D} = 0.2 \, We^{-0.6} \left(\frac{\mu_d}{\mu_c}\right)^{0.07} \tag{5.28}$$

However, in general, theoretical models for average drop size in liquid–liquid, two-phase systems are less frequently used in in-line mixers. A purely experimental approach, typically valid for a very narrow range of physical properties of both phases and for the range of processing conditions (flow rate and rotor speed) is more common. For a small scale, in-line Silverson mixer and for a high volume fraction of very viscous dispersed phases, Gingras et al. [28] proposed the following correlation for the Sauter mean diameter:

$$d_{32} = 0.29 \left(\frac{N^3}{Q}\right)^{-0.55} \phi^{4.13} \mu_d^{-0.53} \tag{5.29}$$

Similar types of correlations for average drop size in rotor–stator mixers were also developed by Averbukh et al. [54] and Koglin et al. [55], but in both cases, the physical properties of the investigated systems were not given; therefore, the applicability of those correlations is rather limited and they are not discussed here.

An alternative description of average drop size in high-energy emulsifying devices such as rotor–stator mixers, colloid mills, or high-pressure homogenizers was developed by Karbstein and Schubert [1]. They related Sauter mean diameter to volumetric specific energy, E_V, defined as the product of energy dissipation rate (describing disruptive stresses) and residence time (the time that drops are exposed to disruptive stresses) [1]:

$$E_V = \frac{P}{Q} = \varepsilon\tau \tag{5.30}$$

and proposed a simple, power-type correlation for average drop size:

$$d_{32} = C_8 E_V^{-b} \tag{5.31}$$

The experimental constants C_8 and b depend on the type of emulsifying device and mechanisms of drop breakage. For in-line rotor–stator mixers, if the drop size is controlled by inertial turbulent breakage, b is of the order 0.35, and it is of the order of 0.75, if drop size is controlled by turbulent shear [3].

5.3.7
Scaling-up of Rotor–Stator Mixers

The majority of work on the power draw and average drop size discussed above were carried out at a very small scale (batch mixers of diameter between 0.025 and 0.05 m) and were aimed at gaining a better understanding of breakage and coalescence in rotor–stator mixers. However, the ultimate aim of all the research was to develop scientifically based scaling-up rules enabling fast and efficient transfer of new, emulsion-based products from the lab to the market.

At present, process development of emulsion-based products often occurs by trial and error to determine the optimum operating conditions to manufacture the desired product. This approach results in high development costs, lost time to market, and considerable material waste due to numerous trials required at ascending sizes from the laboratory scale to plant scale, and final trials in the factory. Appropriate scale-up rules reduce the number of costly experiments necessary to develop a new product, and those rules can also be applied further to future product innovations. This approach of generating scaling rules enables industry to meet new product challenges that are harder to manufacture, have shorter innovation and product life cycles, require improved sustainability of manufacturing processes, and enable increased return on capital investments [56].

There is a large body of work aimed at the development of scale-up rules for emulsification/dispersion processes carried out in stirred vessels. In principle, two scaling parameters are generally considered – the average energy dissipation rate and impeller tip speed. For different impeller/vessel configurations, most of the models based on average energy dissipation rate (Eq. (5.13) or (5.25)) summarized by Leng and Calabrese [57] are moderately successful at different scales. However, it has been also suggested that average drop size during dispersion/emulsification in a stirred vessel scales up well with impeller tip speed, U_T, [58, 59]:

$$d_{32} = C_9 U_T^{-b} \tag{5.32}$$

The literature on scaling-up of rotor–stator mixers is very limited and has been briefly discussed in Section 5.3.5. It is noticeable that practically all the work was

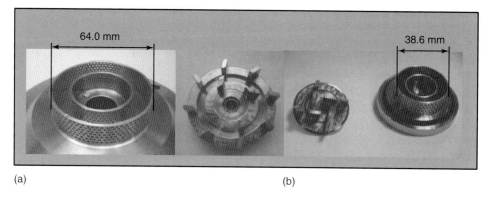

Figure 5.9 Rotors and stators (double-emulsor screen) of "pilot plant, large" (a) and "lab, small" (b) Silverson mixers [12].

carried out with small, batch mixers and, until very recently, there was little in open literature on emulsification/dispersion in in-line rotor–stator mixers. There seem to be two main reasons for the lack of fundamental, systematic investigations of emulsification by in-line rotor–stator mixers. First, until very recently, small in-line mixers suitable for lab applications were not available; secondly, experiments with the mixers of larger sizes are very difficult, expensive, and require large volumes of liquid.

Recently Hall et al. [12] systematically investigated the effect of scale on power draw and DSDs in geometrically similar Silverson rotor–stator mixers shown in Figure 5.9.

DSDs of 9.4 and 339 mPa·s silicone oil in water containing 0.5 wt% sodium laureth sulfate and power draw were measured over wide range of energy dissipation rates and total flow rates. The power draw was measured by calorimetry as this method is relatively simple, and compared to methods based on torque measurements, does not require modification of the mixers as only inlet and outlet temperatures of the emulsions are measured. The power constants in Eq. (5.22) obtained for each scale are practically the same with $Po_Z = 0.23$, $k_1 = 7.5$ in the large scale and $Po_Z = 0.25$, $k_1 = 9.6$ in the small scale, which confirms that the modified expression for power draw in in-line mixers can be used at different scales.

DSDs in the samples withdrawn at the outlet were measured by a Malvern Mastersizer 2000 particle analyzer and typical examples for different rotor tip speeds and different viscosity of the dispersed phase are shown in Figure 5.10.

Figure 5.10 clearly shows that the shapes of DSDs are practically the same at both scales. In the case of low-viscosity dispersed phase (Figure 5.10a) and higher tip speed, distributions are practically lognormal; however, as the speed is reduced, not only are distributions at both scales shifted toward large drops but a tail of small drops appears. Since flow is turbulent in both cases, it appears that the mechanisms of drop breakage also depends on the viscosity of the dispersed phase, which might indicate that shear breakage leading to tip streaming also plays a part in dispersion

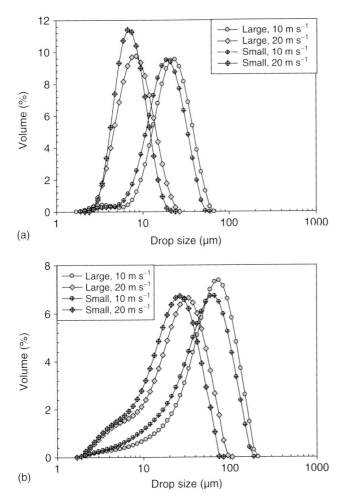

Figure 5.10 Drop size distributions of silicone oils at tip speeds of 10 and 20 m s^{-1} in large and small mixers; (a) 9.4 mPa · s and (b) and 339 mPa · s at the same mean residence time [12].

at lower energy dissipation rates. At high viscosity of dispersed phase, the tail at the lower end of the DSDs is much more pronounced and again reduction of tip speed leads to the reduction of the volume of liquid contained in the smallest drops.

In general, Figure 5.10 shows that there is a very good agreement in the shape of the distributions between the scales and that the maximum drop sizes at both scales are very similar, which indicates that the drop breakage mechanisms are also similar.

In many industrial applications, it is not necessary to describe the efficiency of emulsification in terms of full DSDs and the description in terms of average diameter is sufficient; therefore, the effect of energy dissipation rate and tip speed on Sauter mean diameter (calculated from measured DSDs) has been analyzed and the results are shown in Figure 5.11.

The energy dissipation rate in rotor–stator mixers can be calculated either from power consumption by the rotor (term 1 in Eq. (5.22)) or as a sum of the rotor and flow energy (sum of terms 1 and 2 in Eq. (5.22)). Both cases were analyzed and the difference between the exponents in Eq. (5.14) for the same oil emulsified at different scales was within experimental error. Figure 5.11a showing experimental data (points) and best-fit lines for two viscosities clearly show excellent agreement between the results obtained at different scales, indicating that the breakage mechanism is independent of the scale of the mixers.

However, viscosity of the dispersed phase strongly affects the breakage mechanisms and average drop size. In the case of high-viscosity oil, $b = -0.37$ for both sizes of mixers, practically equal to the theoretical value of $b = -0.4$ for turbulent breakage of drops larger than Kolmogorov's length scale (between 2.0 and 5.8 μm).

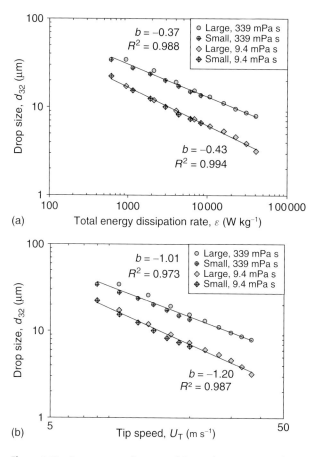

Figure 5.11 Sauter mean diameter of 9.4 and 339 mPa·s silicone oil as a function of (a) total energy dissipation rate and (b) rotor tip speed at two scales [12].

However, for low-viscosity oil, $b = -0.43$ whereas the theoretical value for drops smaller than Kolmogorov's scale $b = -0.5$. This might imply that a substantial part of the oil is in the drops smaller than the Kolmogorov's length scale, which is confirmed by the fact that the Sauter mean diameter at highest energy dissipation rate is of the order of 3 μm, for example, there are many drops smaller than 3 μm.

Plots of dimensionless Sauter diameter against We number are very similar to Figure 5.11a, which is not surprising considering that in all investigated cases interfacial tension was the same; therefore, the ratio of We to energy dissipation rate is constant. Again, all experimental data are very well correlated by a straight line ($R^2 > 0.996$) in log–log coordinates and fall in two distinctive lines described by the following equations:

$$\frac{d_{32}}{D} = 0.29 \, We^{-0.58} \quad \text{for 339 mPa·s} \tag{5.33}$$

$$\frac{d_{32}}{D} = 0.41 \, We^{-0.58} \quad \text{for 9.4 mPa·s} \tag{5.34}$$

Those correlations compare well with correlations developed for stirred tanks by El-Hamouz et al. [59] for similar silicone oil emulsions stabilized by the same surfactant; they found the same trend between the exponent on the We number and viscosity of the dispersed phase.

The effect of tip speed on Sauter mean diameter shown in Figure 5.11b is rather similar to the effect of total energy dissipation rate discussed above. At the same viscosity of the dispersed phase, the Sauter mean diameters measured at each scale also fall in one line in logarithmic coordinates. For high-viscosity oil, $b = -1.01$, and for low-viscosity oil, $b = -1.20$; again, the effect of viscosity on the slope is relatively weak but it strongly affects the intercept.

Considering all the parameters used to scale-up rotor–stator mixers, it appears that at constant residence time, energy dissipation rate is an appropriate scaling parameter for in-line rotor–stator mixers. Impeller tip speed has been recommended for use in scale-up of stirred tanks, but for an in-line rotor–stator mixer rotor tip speed is not the best parameter since the smaller mixer consistently produced slightly smaller drops at equal tip speeds.

One of the frequently asked questions about rotor–stator mixers is whether there is a pronounced difference between the performance of batch and continuous (in-line) devices. The indirect answer to this question is shown in Figure 5.12 where the drop size of two oils measured in in-line rotor–stator mixers at different flow rates are shown.

Figure 5.12 clearly shows that the flow rate has no effect on drop size of very viscous oil and the effect on drop size of low-viscosity oil is not very strong. This result implies that average drop size during emulsification in in-line mixers is effectively determined by the energy coming from the rotor (proportional to N^3), whereas energy associated with the flow (N^2) has only a marginal effect on average drop size. The other important conclusion that can be drawn from Figure 5.12 is that at the same flow rate (Re_h number), the average drop size is practically the same at both scales.

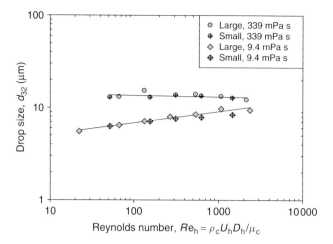

Figure 5.12 The effect of Re_h on drop size of 9.4 and 339 mPa·s silicone oil at a tip speed of 20 m s^{-1} [12].

The last two results have two important practical implications. First, they confirm that at the same geometry of rotor–stator mixer, drop size is independent of the scale. Second, the very weak effect of flow rate on drop size implies that the experiments necessary to scale up in-line rotor–stator mixers can in fact be carried out in batch systems (assuming the geometry of the mixers is similar), which is much easier and cheaper.

5.4
Advanced Analysis of Emulsification/Dispersion Processes in Rotor–Stator Mixers

In the first part of this chapter, the engineering methods based on the concept of average drop size and average energy dissipation rate or shear rate commonly used to design and analyze emulsification processes in many industrial applications have been discussed. Those methods, conceptually very simple, have two major disadvantages. First, they do not allow prediction of DSDs because they assume that the intensity of flow determining drop size (energy dissipation rate or shear rate) is constant in the whole volume of the mixer; second, scaling-up correlations (prediction of power draw and average drop size) are geometry specific.

The crudeness of the assumption of constant flow intensity is obvious from Figure 5.13 where, even for the uninformed reader, it is clear that both intensity and nature of flow in the gap between the rotor and the stator is completely different from the nature and intensity of the flow in the whole of the mixing head volume. Even in much simpler mixing devices such as stirred vessels, it is well accepted that the local energy dissipation rate strongly depends on the position in the vessel, with the ratio of maximum to minimum energy dissipation rate between 80 and 100 [45].

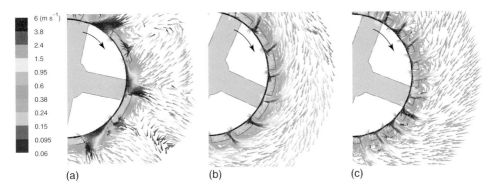

Figure 5.13 Flow patterns around the mixing head fitted with (a) disintegrating screen, (b) slotted screen, and (c) square-hole screen [10].

The fact that scaling-up correlations are geometry specific strongly limits their applicability, and to modify the correlation developed for one type of mixer to another type of rotor–stator mixer, extensive sets of experiments are necessary. This is particularly important during the development of new products, and frequently because of extra cost, manufacturers use "old/existing" mixing devices rather than a device that might be better suited for a new product/process.

The necessity to develop new design methods are forced on manufacturers/researchers by the drive to improve quality and the technical specification of emulsion-based products, something which frequently cannot be accomplished using existing methods. Good examples of such products can be found in the pharmaceutical industry where many emulsion-based drugs require certain DSDs, and also in processes involving multiple emulsions or encapsulation for controlled released of active ingredients.

Detailed discussion of the applicability of advanced methods based on numerical solutions of momentum and mass balances to analyze/design rotor–stator mixers is well outside the scope of this chapter. However, it is the authors' view that the outline of such methods, including examples of selected results, will illustrate their usefulness in analysis of rotor–stator mixers.

Advanced modeling of emulsification involves description of breakage of a single of drop, coalescence between two drops (coalescence between more than two drops is very unlikely and therefore is not considered in engineering applications), and description of the whole population of drops. As all the above phenomena are determined by the flow of the continuous phase, their descriptions require prior knowledge of the velocity/energy dissipation rate distributions inside the rotor–stator mixer.

5.4.1
Velocity and Energy Dissipation Rate in Rotor–Stator Mixers

The velocity and energy dissipation rates distributions in rotor–stator mixers can be calculated using CFD, which involves the solution of momentum balance equations for turbulent flow. The complete set of balance equations including the standard $k - \varepsilon$ turbulence model has the form

$$\frac{\partial U_i}{\partial x_j} = 0 \tag{5.35}$$

$$\frac{\partial U_i}{\partial t} + U_j \frac{\partial U_i}{\partial x_j} = -\frac{1}{\rho^c}\frac{\partial P}{\partial x_i} - \frac{2}{3}\frac{\partial k}{\partial x_i} + (\nu_c + \nu_t)\frac{\partial}{\partial x_j}\left(\frac{\partial U_i}{\partial x_j} + \frac{\partial U_j}{\partial x_i}\right) \tag{5.36}$$

$$\frac{\partial k}{\partial t} + U_j\frac{\partial k}{\partial x_j} = \frac{\partial}{\partial x_j}\left(\nu_c + \frac{\nu_t}{\sigma_k}\right) + \nu_t\left[\left(\frac{\partial U_i}{\partial x_j} + \frac{\partial U_j}{\partial x_i}\right)\frac{\partial U_i}{\partial x_j}\right] - \varepsilon \tag{5.37}$$

$$\frac{\partial \varepsilon}{\partial t} + U_j\frac{\partial \varepsilon}{\partial x_j} = \frac{\partial}{\partial x_j}\left(\nu_c + \frac{\nu_t}{\sigma_\varepsilon}\frac{\partial \varepsilon}{\partial x_j}\right) + C_{\varepsilon 1}\nu_t\frac{\varepsilon}{k}\left[\left(\frac{\partial U_i}{\partial x_j} + \frac{\partial U_j}{\partial x_i}\right)\frac{\partial U_i}{\partial x_j}\right]$$
$$- C_{\varepsilon 2}\frac{\varepsilon^2}{k} \tag{5.38}$$

$$\nu_t = C_\mu \frac{k^2}{\varepsilon} \tag{5.39}$$

and $C_\mu = 0.09, \sigma_k = 1, \sigma_\varepsilon = 1.3, C_{\varepsilon 1} = 1.44, C_{\varepsilon 2} = 1.92$ [60].

In the case of laminar flow, the above set of equations is reduced to Eqs. (5.36) and (5.35) with turbulent viscosity $\nu_t = 0$. For both laminar and turbulent flow, Eq. (5.35) is required, which is a vector equation and is equivalent to three scalar equations describing axial, radial, and tangential velocity at any point inside the rotor–stator mixer. In the case of turbulent flow modeled with the $k - \varepsilon$ model, Eqs. (5.37–5.39) enable calculation of local kinetic energy and energy dissipation rate.

The solution of the above set of equations is often called a *numerical simulation* and it is a well-established research tool in many areas of engineering. It has been frequently used to numerically investigate the effect of impeller type/size and speed on flow patterns and distributions of energy dissipation rate for both Newtonian and non-Newtonian liquids in single and multiphase systems in stirred vessels [61].

The numerical simulation of flow in rotor–stator mixers has been reported far less frequently, probably because it is more demanding. First, it is necessary to ensure spatial resolution of the flow at different length scales: in the gap (length scale of the order of hundreds of microns), in the jets (length scale of the order of millimeters), and around the stator (length scale of the order of tens of centimeters). Second, very high rotor speeds require a very short time step and both time and space resolution make numerical simulation of the flow in rotor–stator mixers very demanding.

5.4.1.1 Batch Rotor–Stator Mixers

One of the first numerical simulations of flow in rotor–stator mixers was published by Calabrese *et al.* [62] who used a two-dimensional simplification of real flow, considerably reducing the complexity of the problem. Recently, full three-dimensional simulations of the flow in a lab-scale batch Silverson rotor–stator mixer have been carried out by Utomo *et al.* [9, 10]. They numerically investigated the effect of rotor speed and geometry of the stator on the velocity field and distributions

5.4 Advanced Analysis of Emulsification/Dispersion Processes in Rotor–Stator Mixers

of energy dissipation rate and reported very good agreement between predicted and experimental velocity distributions [10]. The results of this work clearly prove that CFD can give accurate and detailed description of the flow in the mixer, and the results of numerical simulations can be used to explain manufacturers' statements regarding the effect of geometry of the screen on processing of different materials. In Section 5.2 we quoted the Silverson Web site recommending different types of screens for different dispersion processes, ranging from emulsification to solid breakage/dispersion. Those processes are rather different but in both cases emulsification/dispersion are driven by the flow of the continuous phase, so one can expect that by changing the geometry of the screen, the flow in the mixer is also drastically changed. However, until recently, it was impossible to quantify the differences between the screens, and the results summarized below indicate that while there are some differences between the screens, there are also certain similarities [10].

As shown in Figure 5.13, there are obvious differences in the flow patterns around the mixing heads fitted with different screens. In the case of the disintegrating screen (Figure 5.13a), the high-energy jets emerging from the large holes reach far into the bulk of the fluid and induce flow in the same direction as the rotation of the rotor. There is also relatively intensive recirculation between jets, enhancing bulk mixing. In the case of the slotted screen (Figure 5.13b), the intensity of the jets is reduced and flow in the bulk is driven by recirculation loops. The liquid in the bulk is rotating in the opposite direction to the rotor. Finally, with the square-hole screen (Figure 5.13c) the velocity distribution outside the jets is rather uniform and the flow is not affected as much by circulation loops as in the previous case.

Details of the flow of the jets emerging from differently shaped holes are shown in Figure 5.14. There are certain similarities between the flow, and in all cases, only outflow dominated by the radial component of velocity can be observed in approximately 50% of the hole area. In the other 50%, there is a circulation loop with the fluid flowing toward the center of the mixer. In all cases, the velocity reaches a maximum in close proximity to the leading edge of the hole as the rotor approaches it.

Figure 5.15 clearly shows the effect of the geometry on the distribution of energy dissipation rate both inside and in close proximity to the screens. In all cases, inside the screens, the energy dissipation rate is relatively low and uniform, and, in all cases, the maximum energy dissipation occurs in the relatively small volume at the leading edges of the holes. Also energy dissipation rate in the gap between rotor and stator is close to the maximum values. There is however a major difference in the distributions of energy dissipation rate just outside the screens. In the case of the disintegrating screen, practically all of the energy is dissipated in the volume of the jets, whereas outside the jets, the energy dissipation rate is negligible. From an emulsification point of view, this means that drops are broken in a relatively small volume of liquid in the mixer. In the case of the slotted screen, the energy is dissipated more uniformly, but only in the case of the square-hole screen is the energy uniformly distributed across the screen. This very uniform distribution is caused by the large number of relatively small holes rather than their shape. It

156 5 Emulsification in Rotor–Stator Mixers

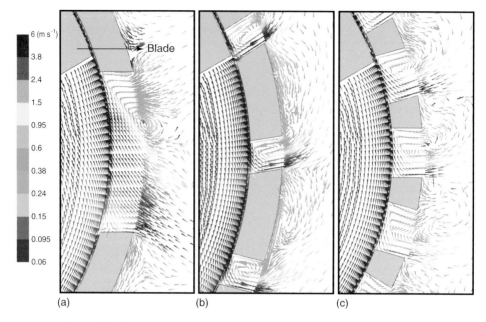

Figure 5.14 Flow patterns of the jets in (a) disintegrating screen, (b) square-hole screen, and (c) slotted screen [10].

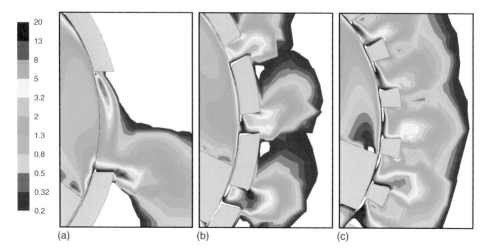

Figure 5.15 Contours of energy dissipation rate (normalized by N^3D^2) for (a) disintegrating head, (b) slotted screen, and (c) square-hole screen. The areas where energy dissipation rate is below 0.2 are shown in white [10].

is necessary to stress here that the results of numerical simulations of the flow just outside the screen were verified experimentally by LDA measurements and calculation of energy dissipation rates were also verified by comparing calculated power number with experimental power number. In both cases, there was very good agreement between theory and experimental results [9].

5.4.1.2 In-Line Rotor–Stator Mixers

The effect of flow rate on velocity and energy dissipation rate distributions inside an in-line Silverson rotor–stator mixer (Figure 5.16) fitted with double-emulsor screens was simulated using the same code but with 2D simplification and a multiple reference frame model being employed. The distributions of radial and tangential velocity at the same rotor speed and two different flow rates in Figure 5.16 show that the effect of flow rate on the velocity field is rather weak. A fourfold increase of flow rate resulted in marginal changes to the velocity field, and the most noticeable difference between the 600 kg h^{-1} (Figure 5.16a) and 2400 kg h^{-1} (Figure 5.16b) cases was slightly higher velocity gradients between the inner and outer screens, and less circulation in the region outside the outer screen. It is interesting to note that the fluid in the region outside the outer screen rotates counterclockwise, whereas both rotors move clockwise. This is similar to the results shown in Figure 5.13b,c for batch mixers fitted with slotted and square-hole screens, which confirms that size/number of holes has a strong effect on the flow in close proximity to the mixing head.

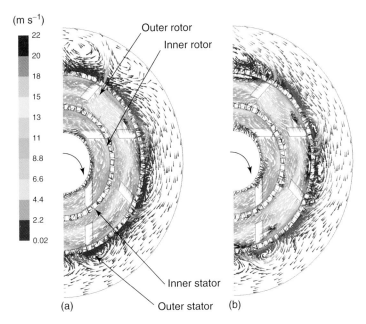

Figure 5.16 Velocity profiles for two flow rates (a) 600 kg h^{-1} and (b) 2400 kg h^{-1}.

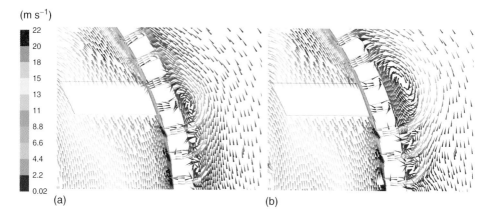

Figure 5.17 Flow patterns of the jets at flow rates of (a) 600 kg h^{-1} and (b) 2400 kg h^{-1}.

Details of the flow in and around the holes are illustrated in Figure 5.17 and again the effect of flow rate is negligible. In both flow rate cases, there is strong counterclockwise rotation just outside the outer screen with some liquid being forced back into the region between the screens. It is interesting to notice the differences of the flow patterns in and around the holes in batch and in-line rotor–stator mixers. In batch mixers (Figure 5.14), the strong recirculation inside all three types of holes was clearly observed but there was no evidence of liquid flowing back into the screens. In in-line mixers (Figure 5.17), large recirculation loops are clearly seen outside the outer screen with liquid flowing back into the region between the screens, through the holes just behind rotor blade. While the flow pattern around the batch mixer was verified experimentally by LDA measurements [10], the results shown for the in-line mixer are only CFD simulation of simplified flows (2D); therefore, full 3D analysis supported by experimental verification is necessary.

The distributions of energy dissipation rate normalized with $N^3 D^2$ at two different flow rates are compared in Figure 5.18. At both flow rates, the energy is mainly dissipated in the region between the inlet and the outer screen, and, in this region, the distributions are rather uniform. As the flow rate increases, the energy dissipation rate in this area marginally increases and at higher flow rates, more energy is dissipated in the jets emerging from the outer screen. Those highly energetic jets emerge only from the holes that are close to the approaching stator blades. It also appears that jets emerging from the inner screen tend to merge forming a layer of fluid close to the screen where energy dissipation rate is high and uniform.

It is noticeable that there are some pronounced differences between the simulated flow in batch and in in-line mixers. The main difference is the character of fluid circulation induced by jets emerging from the holes and the more uniform energy dissipation rate inside the mixers. While in both types of mixers, circulation in the direction opposite to the movement of the rotor was observed, the circulation loops

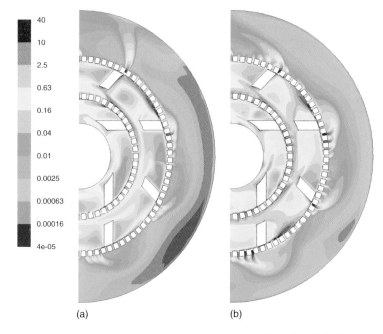

Figure 5.18 Contours of energy dissipation rate (normalized by $N^3 D^2$) for two flow rates (a) 600 kg h^{-1} and (b) 2400 kg h^{-1}.

in batch mixers were much smaller and were confined to the regions adjacent to the jets and holes and no inflow in high-velocity regions was observed. The distributions of energy dissipation rate were also rather different. In batch mixers, high energy dissipation rate is associated with energetic jets and the energy dissipation rate is far from uniform. In in-line, double-screen mixers, the energy is mainly dissipated in the whole region confined by the two screens and the distribution is much more uniform. This difference can be explained by a damping action of the outer screen, which levels out the energy dissipation rate. This also explains why a multiscreen mixer might be more efficient in some emulsification processes than single-screen mixers. It should be noted here that the simulation of flow in in-line double-screen mixers was carried out using 2D approximation of the flow, where only radial and tangential components of velocity were calculated and that these results were not verified experimentally.

Figure 5.19 shows the volumetric distribution of the energy dissipated in the rotor–stator mixer corresponding to the case shown in Figure 5.18a. The distribution of energy is practically bimodal with a high level of energy dissipated inside the stator regions between the inlet and the outer screen, and a relatively low level of energy dissipated outside of the outer screen. There are a few hot spots of very high energy dissipation of the order of 10^6 W kg^{-1} in close proximity to the screens, but these occupy a very small volume of the mixer; therefore, their contribution to emulsification might be questionable.

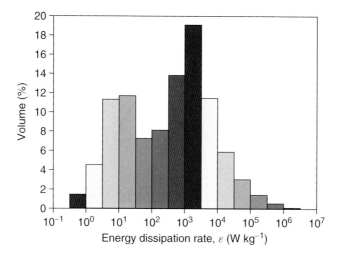

Figure 5.19 Distribution of energy dissipation rate as a function of the rotor–stator mixer volume, $N = 6000$ rpm, $M = 10$ kg min^{-1}.

5.4.2
Prediction of Drop Size Distributions during Emulsification

The DSD in liquid–liquid two-phase systems is determined by the intensity of breakage and coalescence, which occur simultaneously in surfactant-free systems. Both of these phenomena have been extensively investigated theoretically and experimentally, and there are several models relating the frequency of breakage and coalescence to the physical and interfacial properties of both phases, and the type and intensity of the flow of the continuous phase. The majority of those models have been critically reviewed by Liao and Lucas [63, 64]; therefore, only some underlying features of the most commonly used breakage and coalescence models are briefly discussed below.

The breakage model developed by Hinze [43] for a single drop and discussed in Section 5.3 relates drop size to the disruptive and cohesive stresses when those stresses are equal (e.g., at "equilibrium"); this is sufficient to determine the maximum stable drop size in dilute dispersions, but it does not allow description of the kinetics of drop breakage. More advanced models analyze the kinetics of breakage in terms of breakage frequency and number and size distribution of droplets (so-called daughter drops) the large drop is broken into. Those parameters are related to drop size, physical/interfacial properties of both phases, and hydrodynamic conditions quantified by energy dissipation rate in turbulent flow [63] or by shear rate in laminar flow [65]. The aim of such models has been to develop a universal expression for breakage rate that can be used in different systems and at different energy dissipation rates and one of the first models of this type was developed by Coulaloglou and Tavlarides [66]:

$$g(d) = C_{10}d^{-2/3}\frac{\varepsilon^{-1/2}}{1+\phi}\exp\left[-\frac{C_{11}\sigma(1+\phi)^2}{\rho_d\varepsilon^{2/3}d^{5/3}}\right] \qquad (5.40)$$

This expression for breakage rate, as well as similar expressions developed since, practically always contains a certain number (at least one) of unknown constants that have to be calculated from experimental data as discussed further.

Coalescence between two drops has been modeled either as a single-step or a two-step process. Single-step models, also called *empirical*, assume that two drops coalesce immediately as they collide if the energy at collision is sufficiently high [64]. The majority of such models were developed for liquid–liquid dispersion in stirred vessels and coalescence frequency is directly related to impeller speed and diameter, drop size, and volume fraction of the dispersed phase. The major disadvantage of this approach is that the final expression for coalescence frequency is system specific. Two-step models, also called *physical models* [64] postulate that for two drops to coalesce, first, they have to collide and after collision they have to stay in close contact to allow the film of the continuous phase separating them to drain to a certain critical thickness where attractive van der Waals forces dominate and the two drops merge. Therefore, coalescence frequency is described as a product of collision frequency and coalescence efficiency. The collision frequency depends on the drop size and the flow of the continuous phase. The coalescence efficiency depends on the force squeezing drops together therefore it also depends on the flow of the continuous phase and the drop size. Both collision frequency and coalescence efficiency are expressed in terms of energy dissipation rate in turbulent flow and shear rate in laminar flow, drop size, physical properties of both phases and properties of the interface. In turbulent flow, the collision frequency has been modeled assuming that drops follow the random movement of turbulent eddies; therefore, the kinetic theory of gases was frequently used with approach velocity expressed in terms of energy dissipation rate. The coalescence efficiency is typically related to the time necessary for the film separating two drops to drain to a certain critical thickness at which the drops merge. The film drainage time mainly depends upon the magnitude of the forces squeezing two drops together and the type of the interface; rigid, semimobile, or fully mobile [67]. For different types of flow and different types of interface, expressions for collision frequency and coalescence efficiency were developed and were recently reviewed by Liao and Lucas [64]. These expressions are more complex than the expression describing kinetics of breakage and also contain at least two experimental constants. In the great majority of industrial emulsification processes, because coalescence is highly undesirable, it is prevented by the selection of appropriate surfactants; details of coalescence are not discussed here.

Breakage and coalescence models are used within the framework of population balance modeling (mass balance for dispersed systems) to calculate DSDs from the solution of the following equation [68]:

$$\frac{\partial n(v,t)}{\partial t} + \nabla\left[U_e n(v,t)\right] + \nabla\left[U n(v,t)\right] = B - D \qquad (5.41)$$

The first term on the left describes the time change of a number of drops of size v, the second term describes the net change of drops of size v due to continuous growth by mass transfer, and the third term describes the change of number of drops of size v by convection. The terms on the right describe disappearance of drops (death, D) and appearance of drops (birth, B) which result from breakage and coalescence. This general version of the population balance can be simplified for the majority of emulsification processes by neglecting both the term describing continuous reduction of drop size (due to dissolution in the systems with partial mutual solubility) and coalescence:

$$\frac{\partial n(v,t)}{\partial t} + \nabla [Un(v,t)] = -g(v)n(v,t) + \int_v^\infty \beta(v,v')\nu(v')g(v')n(v',t)dv'$$

(5.42)

where, $n(v,t)$ is the number of drops with size v per unit volume, $\beta(v,v')$ is the size distribution of daughter droplets, $g(v)$ is the breakage rate of drops of size v discussed above, and $\nu(v')$ is the number of daughter drops formed by breakage of a drop of size v'. Typically the number of daughter drops is equal to two and the functions used to describe their distribution are summarized by Liao and Lucas [64]. Equation (5.42) can be further simplified by assuming that the system is perfectly mixed; therefore, all the spatial gradients are equal to zero and the second term on the right-hand side of Eq. (5.42) equals zero. The last assumption holds reasonably well for dispersion/emulsification in stirred vessels and simplified versions of population balance models have been frequently solved for different types of dispersed phase systems such as gas bubbles suspended in liquid, solid particles suspended in liquid, mixtures of two immiscible liquids, and also for drops or particles suspended in gas.

However, simultaneous solutions of momentum balance for three-dimensional turbulent flow (Eqs. 5.36–5.39) and population balance equations (Eq. (5.42)) is far from trivial and only relatively recently Baldyga et al. [14] used this approach to analyze breakage of aggregates formed from nanoparticles. They solved population balance equations describing the kinetics of breakage of aggregates using the quadrature methods of moments [69] linked with the solver of momentum balance equations (Eqs. 5.36–5.39) within FLUENT. The calculated transient size distributions during breakage were in good agreement with experimental data.

Clearly the advanced methods of analysis of flow and/or emulsification in in-line rotor–stator mixers offer a unique tool enabling a better understanding of the processes occurring inside these devices of rather complex geometry and operating at very high rotor speeds. The solution of momentum balance equations has been verified experimentally by LDA [9] and it has been accepted that, in general, there is a good agreement between theory and experiments.

Numerical simulations are attractive for industry as they can significantly reduce the number of experimental trials required to develop a new product, and can be used as a virtual manufacturing tool to accelerate new equipment design. The reduction of experimental trials is beneficial since they are time consuming, labor intensive, produce waste material, and are expensive. However, despite the benefits of numerical simulations, new formulations and equipment designs will also need to be verified experimentally, albeit with the number of experiments drastically reduced.

5.5 Conclusion

Rotor–stator mixers have been widely used in industry to manufacture a range of emulsion/dispersion based products over many years as they provide a relatively simple and effective way of providing very intense mixing and dispersion. They are also attractive for industry as they are considered to be relatively inexpensive, robust, and versatile; therefore, they can be used in a variety of applications in the food, personal care, pharmaceutical, and fine chemical industries. Rotor–stator mixers are available in a wide range of designs that can be tuned to fulfill requirements of different processes and they are suitable for batch (small-scale) and continuous (large-scale) processing.

Their application could be further extended and optimized if the fundamental knowledge and understanding of the effects of their design, processing conditions, and physical/interfacial properties of the processed material on the quality of the products were improved. In this chapter, we summarized the existing, rather basic description of the flow and dispersion processes currently used in engineering applications, and also pointed out the areas particularly suitable for advanced research/design. Definitely, CFD offers unique possibilities in describing single-phase flow; however, even in this case, many would argue that at least partial experimental verification of the theoretical results is necessary. Application of CFD in multiphase systems such as liquid–liquid (emulsification) or solid–liquid (deagglomeration, dispersion) is more challenging as detailed modeling of the interactions in such systems is still very limited. However, development of new breakage and coalescence models within the population balance equations combined within CFD and verified by experiments should enable more detailed analysis of emulsification processes in rotor–stator mixers.

Nomenclature

a	Specific interfacial area	m^2 m^{-3}
B	Birth rate of drops	s^{-1}
b	Exponent	—
$C_{1\ldots 11}$	Empirical constants	—
$C_{\varepsilon 1}$	Constant in standard $k-\varepsilon$ model	—
$C_{\varepsilon 2}$	Constant in standard $k-\varepsilon$ model	—
C_μ	Constant in standard $k-\varepsilon$ model	—
D	Characteristic diameter (outer rotor diameter)	m

D	Death rate of drops	s^{-1}
d	Drop diameter	m
D_h	Drop diameter of stator holes	m
R^2	Linear regression correlation coefficient	—
d_{10}	Number ean diameter	m
d_{32}	Sauter (volume surface) mean diameter	m
d_{43}	Mass mean diameter	m
d_{max}	Maximum stable drop diameter	m
E_V	Energy density	$J\,m^{-3}$
$g(d)g(v), g(v')$	Breakage rate of drops of size d, v, v'	s^{-1}
K	Proportionality constant	—
K_1	Shear constant	—
k_0	Laminar flow constant	—
k	Turbulent kinetic energy per unit mass	$m^2\,s^{-2}$
k_1	Power flow constant	—
M	Mass flow rate	$kg\,s^{-1}$
N	Rotor speed	s^{-1}
$n(v, t), n(v, t)$	Number of drops of size v, v per unit volume	—
Q	Volumetric flow rate	$m^3\,s^{-1}$
P	Pressure	Pa
P	Power	W
P_L	Power losses term	W
p	Integer	—
Po_z	Power number constant for in-line rotor–stator mixers zero flow	—
q	Integer	—
r	Radius of droplet	m
t	Time	s
U	Fluid velocity	$m\,s^{-1}$
U_i	Fluid velocity in i direction	$m\,s^{-1}$
U_j	Fluid velocity in j direction	$m\,s^{-1}$
U_e	Internal velocity	$m\,s^{-1}$
U_T	Rotor tip speed	$m\,s^{-1}$
V	Mixing head volume (swept rotor volume)	m^3
x_i, x_y	Cartesian coordinates	—
v	Volume of drops	m^3
y	Displacement	m

Greek symbols

$\beta(v, v')$	Frequency of daughter drops of size v'	—
$\dot{\gamma}$	Shear rate	s^{-1}
δ	Rotor–stator gap width	m
ε	Total energy dissipation rate per unit mass of fluid	$W\,kg^{-1}$
λ	Viscosity ratio	—
μ_c	Continuous phase viscosity	Pa s
μ_d	Dispersed phase viscosity	Pa s
$v(v')$	Number of daughter drops formed by breakage of a drop of size v'	—
v_c	Kinematic viscosity	$m^2\,s^3$
v_t	Turbulent viscosity	$m^2\,s^{-3}$
ρ_c	Continuous phase density	$kg\,m^{-3}$
ρ_d	Dispersed phase density	$kg\,m^{-3}$

σ	Interfacial tension	N m^{-1}
σ_ε	Constant in standard k–ε model	—
σ_k	Constant in standard k–ε model	—
τ	Residence time	s
τ_c	Cohesive stresses acting on a droplet	N m^{-2}
τ_d	Disruptive stresses acting on a droplet	N m^{-2}
ϕ	Volume fraction of dispersed phase	—

Dimensionless groups

Ca	Capillary number $\frac{\dot{\gamma}\mu_c d}{\sigma}$	—
Ca_{cr}	Critical capillary number $\frac{\dot{\gamma}\mu_c d_{max}}{\sigma}$	—
Po	Power number $\frac{P}{\rho N^3 D^5}$	—
Re	Reynolds number $\frac{\rho_c N D^2}{\mu_c}$	—
Re_h	Reynolds number in stator holes $\frac{\rho_c U_h D_h}{\mu_c}$	—
We	Weber number $\frac{\rho_c N^2 D^3}{\sigma}$	—

References

1. Karbstein, H. and Schubert, H. (1995) *Chem. Eng. Proc.*, **34**, 205–211.
2. Atiemo-Obeng, V.A. and Calabrese, R.V. (2004) in *Handbook of Industrial Mixing* (eds E.L. Paul, V.A. Atiemo-Obeng, and S.M. Kresta), John Wiley & Sons, Inc., New York, NY, pp. 479–505.
3. Schubert, H. and Engel, R. (2004) *Chem. Eng. Res. Des.*, **82**, 1137–1143.
4. Pacek, A.W., Ding, P., and Utomo, A.T. (2007) *Powder Technol.*, **173**, 203–210.
5. Ding, P., Orwa, M.G., and Pacek, A.W. (2009) *Powder Technol.*, **195**, 221–226.
6. Muller-Fischer, N., Suppiger, D., and Windhab, E.J. (2007) *J. Food Eng.*, **80**, 306–316.
7. Utomo, A.T., Baker, M., and Pacek, A.W. (2009) *Chem. Eng. Res. Des.*, **87**, 533–542.
8. Becher, P. (ed.) (1985) *Encyclopaedia of Emulsion Technology*, Marcel Dekker, Inc., New York, NY.
9. Utomo, A.T., Baker, M., and Pacek, A.W. (2008) *Chem. Eng. Res. Des.*, **86**, 1397–1409.
10. Utomo, A.T. (2009) Flow patterns and energy dissipation rates in batch rotor–stator mixers. PhD thesis. University of Birmingham, UK.
11. Hall, S., Cooke, M., El-Hamouz, A., and Kowalski, A.J. (2011a) *Chem. Eng. Sci.*, **66**, 2068–2079.
12. Hall, S., Pacek, A.W., Kowalski, A.J., Cooke, M., and Rothman, D. (2011b) *Can. J. Chem. Eng.*, **89** (5), 1040–1050.
13. Baldyga, J., Kowalski, A., Cooke, M., and Jasinska, M. (2007) *Chem. Process Eng.*, **28**, 867–877.
14. Baldyga, J., Orciuch, W., Makowski, L., Malik, K., Ozcan-Taskin, G., Eagles, W., and Padron, G. (2008) *Ind. Eng. Chem. Res.*, **47**, 3642–3663.
15. http://www.silverson.co.uk/en/products.html (accessed 30 September 2011).
16. http://www.highshearmixers.com/models.html (accessed 30 September 2011).
17. http://www.chemineer.com/greerco-_high_shear_applications.php (accessed 30 September 2011).
18. http://www.ikausa.com/rotor-stator.htm (accessed 30 September 2011).
19. http://www.ytron-quadro.co.uk/details.asp?ProdID=6 (accessed 30 September 2011).
20. http://www.siefer-trigonal.de/machines (accessed 30 September 2011).
21. http://www.klausen.net.au/agencies/17-siefer (accessed September 30 2011)

22. Davies, J.T. (1987) *Chem. Eng. Sci.*, **42** (7), 1671–1676.
23. Myers, K.J., Reeder, M.F., Ryan, D., and Daly, G. (1999) *Chem. Eng. Prog.*, **95**, 33–42.
24. Deutsch, D. (1998) *Chem. Eng.*, **105** (8), 76.
25. Davies, J.T. (1985) *Chem. Eng. Sci.*, **40** (5), 839–842.
26. Wieringa, J.A., Van Dieren, F., Janssen, J.J.M., and Agterof, W.G.M. (1996) *Trans. IChemE*, **74** (A), 554–562.
27. Brocart, B., Tanguy, P.A., Magnin, C., and Bousquet, J. (2002) *J. Dispersion Sci. Technol.*, **23**, 45–53.
28. Gingras, J.-P., Tanguy, P.A., Mariotti, S., and Chaverot, P. (2005) *Chem. Eng. Proc.*, **44**, 979–986.
29. Bengoechea, C., Lopez, M.L., Cordobes, F., and Guerrero, A. (2009) *Food Sci. Technol. Int.*, **15**, 367–373.
30. Formiga, F.R., Fonseca, I.A.A., Souza, K.B., Silva, A.K.A., Macedo, J.P.F., Araujo, I.B., Soares, L.A.L., Socrates, E., and Egito, T. (2007) *Int. J. Pharm.*, **344**, 158–160.
31. El-Jaby, U., McKenna, T.F.L., and Cunningham, M.F. (2007) *Macromol. Symp.*, **259**, 1–9.
32. Maa, Y.-F. and Hsu, C. (1996) *J. Controlled Release*, **38**, 219–228.
33. Masmoudi, H., Le Dreau, Y., Piccerelle, P., and Kister, J. (2005) *Int. J. Pharm.*, **289**, 117–131.
34. Yuan, Q., Williams, R.A., and Biggs, S. (2009) *Colloids Surf. A*, **347**, 97–103.
35. Fradette, L., Brocart, B., and Tanguy, P.A. (2007) *Chem. Eng. Res. Des.*, **85** (A11), 1553–1560.
36. Jacquier, E. and Jacquier, J.C. (2009) *J. Food Eng.*, **94** (3–4), 316–320.
37. Perrier-Cornet, J.M., Marie, P., and Gervais, P. (2005) *J. Food Eng.*, **66**, 211–217.
38. Kowalski, A.J. (2009) *Chem. Eng. Process.*, **48**, 581–585.
39. Rothman, D. (2009) Mixing for formulated products. IChemE FPE Meeting.
40. Ding, P. and Pacek, A.W. (2008) *J. Colloid Interface Sci.*, **325**, 165–172.
41. Calabrese, R.V., Francis, M.K., Kevala, K.R., Mishra, V.P., and Phongikaroon, S. (2000) in *10th European Conference on Mixing* (eds H.E.A. van den Akker and J.J. Deksen), Elsevier Science, Amsterdam, pp. 149–156.
42. Walstra, P. and Smulders, P.E.A. (1998) Emulsion formation, in *Modern Aspects of Emulsion Science* (ed. B.P. Binks), The Royal Society of Chemistry.
43. Hinze, J.O. (1955) *AIChE J.*, **1**, 289–295.
44. Baldyga, J. and Bourne, J.R. (1999) *Turbulent Mixing and Chemical Reactions*, John Wiley & Sons, Inc.
45. Edwards, M.F., Baker, M.R., and Godfrey, J.C. (1997) in *Mixing in the Process Industries*, 2nd edn (eds N. Harnby, M.F. Edwards, and A.W. Nienow), Butterworth Heinemann, Oxford, pp. 137–158.
46. Doucet, L., Ascanio, G., and Tanguy, P.A. (2005) *Trans. IChemE A*, **83**, 1186–1195.
47. Padron, G.A. (2001) Measurement and comparison of power draw in batch rotor–stator mixers, MSc thesis. University of Maryland, College Park, MD.
48. Sparks, T. (1996) Fluid mixing in rotor–stator mixers. PhD thesis. Cranfield University, UK.
49. Kowalski, A.J., Cooke, M., and Hall, S. (2011) *Chem. Eng. Proc.*, **66**, 241–249.
50. Cooke, M., Rodgers, T.L., and Kowalski, A.J. (2012) Power consumption characteristics of an in-line Silverson high shear mixer, *AIChE J.*, **58** (6), 1683–1692.
51. Pacek, A.W., Chamsart, S., Nienow, A.W., and Bakker, A. (1999) *Chem. Eng. Sci.*, **54**, 4211–4222.
52. Baldyga, J., Bourne, J.R., Pacek, A.W., Amanullah, A., and Nienow, A.W. (2001) *Chem. Eng. Sci.*, **56**, 3377–3385.
53. Puel, F., Briancon, S., and Fessi, H. (2006) in *Microencapsulation: Methods and Industrial Applications*, 2nd edn, vol. 158 (ed. S. Benita), Drugs and Pharmaceutical Science, Chapter 6, CRC Press Taylor & Francis Group, pp. 149–182.
54. Averbukh, Yu.N., Nikoforov, A.O., Kostin, N.M., and Korshakov, A.V. (1988) *J. Appl. Chem. USSR*, **61** (2), 396–397.
55. Koglin, B., Pawlowski, J., and Schnoring, H. (1981) *Chem.-Ing.-Technol.*, **53** (8), 641–647.

56. Kowalski, A.J., Hill, M., Marshman, C., and Pearson, C. (2005) Challenges facing manufacture of complex structured products. APACT 05, Stratford-upon-Avon, UK.
57. Leng, D.E. and Calabrese, R.V. (2004) in *Handbook of Industrial Mixing* (eds E.L. Paul, V.A. Atiemo-Obeng, and S.M. Kresta), John Wiley & Sons, Inc., New York, pp. 639–753.
58. Okufi, S., Perez de Ortiz, E.S., and Sawistowski, H. (1990) *Can. J. Chem. Eng.*, **68**, 400–406.
59. El-Hamouz, A., Cooke, M., Kowalski, A., and Sharratt, P. (2009) *Chem. Eng. Proc.*, **42** (2), 633–642.
60. Fluent Inc. (2006) User's Guide, Fluent 6.3.
61. Joshi, J.B., Nere, N.K., Rane, C.V., Murthy, B.N., and Mathpati, C.S. (2011) *Can. J. Chem. Eng.*, **89**, 754–816.
62. Calabrese, R.V., Francis, M.K., Kevala, K.R., Mishra, V.P., Padron, G.P., and Phongikaroon, S. (2002) Fluid dynamics and emulsification in high shear mixers. Proceedings of the 3rd World Congress on Emulsions, Lyon, France.
63. Liao, Y. and Lucas, D. (2009) *Chem. Eng. Sci.*, **64**, 3389–3406.
64. Liao, Y. and Lucas, D. (2010) *Chem. Eng. Sci.*, **65**, 2851–2864.
65. Bentley, B.J. and Leal, L.G. (1986) *J. Fluid Mech.*, **167**, 219–240.
66. Coulaloglou, C.A. and Tavlarides, L.L. (1976) *AIChE J.*, **22** (2), 289–297.
67. Chesters, A.K. (1991) *Trans. IChemE*, **69A**, 259–270.
68. Ramkrishna, D. (2000) *Population Balances, Theory and Application to Particulate Systems in Engineering*, Academic Press, San Diego.
69. Marchisio, D.L., Vigil, R.D., and Fox, R.O. (2003) *J. Colloid Interface Sci.*, **258**, 322–334.

6
Formulation, Characterization, and Property Control of Paraffin Emulsions
Jordi Esquena and Jon Vilasau

6.1
Introduction

Paraffin (also referred to as *paraffin wax*) derivates directly from slack wax, a by-product of refineries (petroleum cracking) for preparation of lubrication oils. The main difference between paraffin emulsions and other types of emulsions is that paraffin is usually solid at room temperature. Therefore, the emulsions are prepared at high temperature, above the melting point of the paraffin. Once the emulsion is formed, the temperature is rapidly decreased and paraffin is solidified. Thus, at room temperature paraffin emulsions become particle dispersions.

Paraffin emulsions have already been used in industry since long ago, in a wide variety of different technological applications. Industrial formulators have achieved a remarkably good control of particle size and stability, and optimum formulations are being produced, which target distinctive applications. However, these formulations have often been reached by using a trial-and-error approach, without making use of fundamental background in colloid and interface science.

Actually, many aspects in the industrial performance of paraffin emulsions can be explained by applying known theoretical principles, which may allow predicting the performance of commercial products. However, understanding practical problems in the industrial world is often difficult, due to formulation with impure components, imprecise characterization of physicochemical properties and unknown process parameters.

In this chapter, the aim is to provide an insight on paraffin emulsions that are commonly used, but often not well understood from a scientific point of view. Herein, systematic studies on formulation, stabilization, and properties of paraffin emulsions are presented. This chapter reviews recent works published in a series of papers [1–6], which focused on paraffin emulsions. However, it should be pointed out that the same approach and methodology can be applied to study other types of emulsions, as paraffin is not fundamentally different from other types of oils with high melting points.

Emulsion Formation and Stability, First Edition. Edited by Tharwat F. Tadros.
© 2013 Wiley-VCH Verlag GmbH & Co. KGaA. Published 2013 by Wiley-VCH Verlag GmbH & Co. KGaA.

6.1.1
Industrial Applications of Paraffin Emulsions

Paraffin emulsions, commonly known in industry as wax emulsions, are used in a wide range of technological applications such as coating in the food packaging industry or providing waterproof properties to particleboard panels in the furniture industry. Regarding the boards, the increased market demand for logwood in pulp and paper industries, as well as the increasing use of wood pellets has led to high costs and shortage in the supply of wood-derivative-processed products [7]. For this reason, the use of particleboards made from vegetal fibers from plants (sugar cane, vine prunings, flax shiv, etc.) or from fruit waste (banana, coconut, date palm, peanut, kiwi, pineapple, etc.) is increasing significantly. These types of boards are commonly known as *bagasse particleboard* and could be a real alternative to substitute the wood-based panels [8–15]. Particleboard panels involve any kind of board made from either wood shavings or bagasse fibers using a binder in its elaboration. Paraffin emulsion plays a key role in this sector as an additive, improving waterproof properties. The interest in this field has increased and it requires a deep understanding of the physicochemical properties of paraffin emulsions.

In industry, it is of prime importance to formulate the emulsions appropriately, reducing costs of formulation, controlling properties such as particle size and enhancing stability (against shear, freeze–thaw cycles, and electrolyte concentration). This chapter focuses on these aspects: discussing about the selection of surfactants, how to control the droplet size, and achieve the desired stabilization. In the next sections, some general aspects about paraffin and paraffin emulsions are described.

6.1.2
Properties of Paraffin

Paraffin consists of a mixture of saturated linear hydrocarbons with an average chain length between 20 and 40 carbon atoms and an average melting point between 40 and 70 °C. Its waterproof properties and chemical inertia make paraffin suitable for many different applications such as candles, container coatings, or impregnating paper [16–18].

One of the most important properties of paraffin is its crystalline nanostructure. Depending on the temperature, paraffin chains can adopt different arrangements and usually its characterization is performed by means of X-ray dispersion techniques such as wide angle X-ray scattering (WAXS). As illustration, Figure 6.1 shows the X-ray scattering spectra of paraffin. In this example, the paraffin consisted of a mixture of saturated hydrocarbons with an average chain length of 25 carbon atoms and a melting point of 52 °C, measured by DSC (differential scanning calorimetry).

At 25 °C, the spectrum of paraffin shows a great number of very sharp diffraction peaks, which can be observed at both small ($q < 5$ nm^{-1}) and wide angles ($q > 10$ nm^{-1}). These peaks indicate a high degree of crystallinity. At small angle,

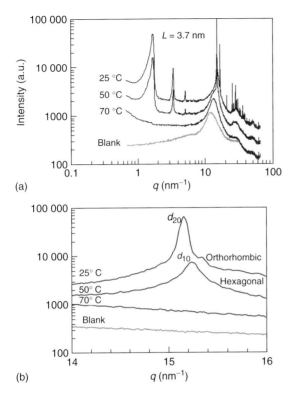

Figure 6.1 X-ray spectra of the paraffin as a function of temperature. For clarity, intensity was multiplied by an arbitrary constant (arbitrary units, a.u.). (a) q-range from 0.8 to 60 nm^{-1} and (b) q-range from 12 to 18 nm^{-1}.

the three peaks correspond to a lamellar symmetry with a repeat distance of 3.7 nm (indicated as L in Figure 6.1a), in a similar packing as smectic phases. At wide angle, the peak with the highest intensity ($d_{20} = 0.414$ nm, in Figure 6.1b) provides information about the packing of the alkyl chains (Figure 6.2). Typically, solid crystalline alkanes show an intense peak at approximately 0.41 nm, which is also observed in many other organic compounds with crystalline alkyl chains, such as in surfactants forming gel phases (L_β) [19, 20]. It is known [21] that the alkyl chains may pack in different structures; mainly orthorhombic or hexagonal (Figure 6.2). Focusing the attention in a narrower q-range of the spectra (14–16 nm^{-1}), the shoulder observed next to the d_{20} sharp peak indicates a crystalline structure with orthorhombic geometry [21]. Therefore, at 25 °C, the linear molecules pack forming lamellar structures that consist of orthorhombic cells, as presented in Figure 6.2b.

At 50 °C, the degree of ordering (crystallinity) of the paraffin has decreased, as indicated by lower peak intensity. At small angle, the same three peaks are observed demonstrating that the lamellar structure is also present, and the 3.7 nm distance is not varied. However, at wide angle, only the most intense peak ($d_{10} = 0.412$ nm) is observed, and the shoulder vanishes (Figure 6.1b) indicating that a transition

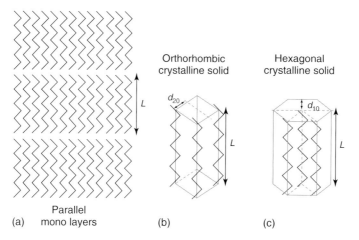

Figure 6.2 Crystalline structure of a paraffin. Organization of the chains in (a) parallel monolayers, (b) orthorhombic geometry, and (c) hexagonal geometry.

from orthorhombic to hexagonal structure took place (Figure 6.2b and 6.2c) [21]. Consequently, the paraffin chains are still in a crystalline solid state at 50 °C, but possess certain mobility associated with the hexagonal structure [22].

At 70 °C, above the paraffin melting point, the molecules become highly disordered, leading to an amorphous liquid, as indicated by the absence of diffraction peaks at both small and wide angles (Figure 6.1). The broad band that is observed at wide angle is due to the glass capillary, where measurements were performed, as it was also observed in the blank. In summary, the paraffin showed transitions from orthorhombic to hexagonal, and finally to amorphous liquid, when increasing temperature.

In any case, metastable structures may be formed, especially when fast cooling (quenching) is produced. Consequently, kinetically controlled recrystallization processes can be observed, which lead to the formation of the most stable structure at a given temperature. A typical example is the slow change in texture observed in some candles made of paraffin.

6.1.3
Preparation of Paraffin Emulsions

As paraffin is solid at room temperature and possesses high viscosity above its melting point, plugging or difficult pumping can occur during circulation through pipelines. Consequently, emulsification of paraffin was a logical solution to prevent these industrial problems [23].

The basic principles explained next are not only valid for paraffin emulsions but also for emulsions in a broad sense. It is well known that emulsions are not spontaneously formed and the free energy of formation can be easily calculated from

$$\Delta G = \gamma \Delta A - T \Delta S \tag{6.1}$$

where ΔG represents the Gibbs free energy; γ, the interfacial tension between two liquids; ΔA, the increase in interfacial area; ΔS, the increase in entropy; and T, the temperature. As $\gamma \Delta A > T \Delta S$, the energy of formation will be always positive, which implies nonspontaneous formation and requirement of energy.

Depending on the nature of this energy input, emulsion formation methods can be classified as low-energy methods (condensation methods) and high-energy methods (dispersion methods). In the former, the emulsification is induced by phase transitions produced by changes in temperature and/or composition, and the energy input comes from the chemical potential of the system. On the other hand, high-energy methods make use of external devices such as rotor–stator, ultrasonic, or high-pressure systems to apply energy by shear. The emulsification is usually produced by deformation of droplets under shear. Droplet disruption is opposed by Laplace pressure, which for monodisperse systems takes the following form

$$\Delta P = (P_{int} - P_{ext}) = \frac{2\gamma}{r} \tag{6.2}$$

where P_{int} and P_{ext} represent pressures inside and outside the droplet, respectively, and r the droplet radius. Emulsification takes place when the oil–water interface is deformed, requiring an applied pressure higher than Laplace pressure [24]. Because of that, emulsification by dispersion methods usually requires an energy input much higher than the thermodynamic free energy of formation ΔG. Often, a great amount of energy is required to overcome high-energy barriers during emulsification and actually most of the energy is lost because of viscous dissipation.

In order to obtain stable paraffin emulsions, small particle size (about 1.0 μm) and low polydispersity are required. In industry, such properties are generally achieved using high-energy methods, mainly based on high-pressure homogenization, as the high melting point of paraffin makes it difficult to use low-energy methods. Therefore, emulsification is commonly achieved by applying high shear above the melting point of the paraffin, and the Krafft point of the surfactants. Afterward, emulsions are rapidly cooled down to prevent destabilization and surfactant desorption. Traditionally, in industry, paraffin emulsions have been stabilized with an ionic surfactant that produces long-range electrostatic repulsion [23].

However, recent works in the field of paraffin emulsions have revealed that the use of ionic/nonionic mixed surfactant systems provides electrosteric repulsion enhancing emulsion stability [1–3]. At high ionic/nonionic surfactant ratios, a significant change in emulsion properties was observed. The increase in the storage modulus (G') and the decrease in particle size led to an increase in emulsion stability. In addition, phase behavior of the surfactant system in water revealed that lamellar liquid crystalline aggregates were present. It was considered that the high stability could be related to the formation of a surfactant multilayer arrangement surrounding the particles (Figure 6.3), as already explained by other authors [25–30].

Friberg et al. [28] described the stabilization by multilayers around the droplets, as a factor that could increase viscosity and enhance emulsion stability. This

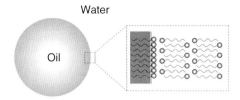

Figure 6.3 Example of possible stabilization of emulsion droplets by a multilayer arrangement of interdigitated surfactant chains. (Reproduced with permission from Ref. [5].)

stabilization mechanism was observed in formulations with a 1 : 1 oil/water ratio and surfactant concentrations between 3 and 4 wt%. Nevertheless, as also shown [31–33], the surfactant concentration required to form multilayers surrounding the droplets is normally higher (between 10 and 15 wt%). Moreover, it should be remarked that the formation of such multilayers depends on many factors such as the nature of the surfactant, the interfacial area generated in the emulsification process, and the surfactant solubility in the continuous phase.

6.2
Surfactant Systems Used in Formulation of Paraffin Emulsions

Mixtures of nonionic and ionic surfactants have been used for stabilizing oil-in-water emulsions because a combination of steric and electrostatic forces is beneficial to enhance stability [34]. Traditionally, anionic surfactants have been used to stabilize paraffin particles. The ionic surfactant usually consisted of an acid, with an alkyl chain length between 17 and 21, neutralized by a strong base [23]. Recent works about paraffin emulsions [5] studied the influence of the addition of a nonionic surfactant on emulsion stability.

It should be remarked that in industry, paraffin emulsions are usually stored during very long periods. Furthermore, these emulsions are often diluted using hard water that contains high-electrolyte concentration. In addition, the emulsions are pumped through pipelines producing high shear that may cause emulsion destabilization. Consequently, the selection of an appropriate ionic/nonionic surfactant mixture is of prime importance, in order to optimize electrosteric repulsion, which will retard destabilization produced by flocculation, creaming, or coalescence [35, 36]. Nonionic surfactant provides steric repulsion, whereas ionic surfactant increases electrostatic repulsion.

In the formulation of any surfactant system, as well as in other aspects of colloidal science, phase behavior is a useful approach [37]. In paraffin emulsions, phase behavior of water/surfactant systems has been thoroughly studied in order to determine possible aggregates that can be present in the emulsions. As mentioned before, a multilayer arrangement could contribute to enhance emulsion stability.

6.2.1
Phase Behavior

An appropriate surfactant system, as described by Mújika-Garai et al. [2] and later by Vilasau et al. [5], consisted of a mixture of ionic and nonionic surfactants. In industry, surfactants with long alkyl chains are commonly used, as their adsorption energy is higher, allowing better stability. Moreover, mixtures of ionic and nonionic surfactant are used to achieve enhanced electrosteric stabilization. However, the Krafft and melting points of these long-chain surfactants as well as the melting point of paraffin may be high and, consequently, temperature should be high enough during emulsification. Therefore, the information about surfactant phase behavior is important, as it allows a correct selection of emulsification temperature.

In the present section, the phase behavior of an ionic and a nonionic surfactant are presented, in which the alkyl chain lengths are relatively long, 21 and 19 carbon atoms, respectively. These surfactants are presented as examples for illustration. The ionic surfactant (anionic) was formed by neutralization of a fatty acid with an excess of alkanolamine. The stoichiometric excess was used to ensure complete neutralization. The nonionic surfactant was a conventional ethoxylated fatty alcohol.

Below the melting points of these surfactants, lamellar interdigitated gel phases are formed. These structures are identified by SAXS (small-angle X-ray scattering) measurements. In the present case, both surfactants showed SAXS spectra with peak sequences associated to lamellar structures, and a sharp peak at high q values (15.3 nm) indicating gel structures (L_β) [38]. Moreover, considering all interatom distances in the molecule, the experimental interlayer distances of both surfactants are slightly shorter than the theoretical molecule length [39]. The alkyl chains would penetrate into each other forming a monolayer with polar groups at both sides.

This gel structure is able to absorb small amounts of water and therefore, the solubility boundary of both ionic and nonionic surfactants is above the melting point of the pure surfactant (Figure 6.4). The nonionic surfactant in water forms lamellar aggregates, (Figure 6.4a) whereas the ionic surfactant forms hexagonal aggregates (Figure 6.4b), at high-surfactant concentration. Probably, this hexagonal structure is inverse (H_2), considering the natural sequence of phases (curvature of aggregates) that surfactants usually form, when increasing their concentration. The self-assembled structures were identified by SAXS and polarized optical microscopy (POM). At low surfactant concentration, different multiphase regions were observed. The *water/nonionic surfactant system* forms lamellar liquid crystal coexisting with aqueous phase ($W + L_\alpha$) down to 1 wt% surfactant. However, The *water/ionic surfactant system* forms liquid crystal coexisting with an excess of aqueous phase ($W + LC$). Two different types of liquid crystalline structures were identified in this region. At very low surfactant concentration (down to 0.2 wt%), lamellar structures were identified, whereas at higher surfactant compositions hexagonal structures were observed. Both surfactants present very low water solubility and no single-phase micellar solution was observed. However, probably a single-phase region may exist at very low surfactant concentrations (compositions were studied down

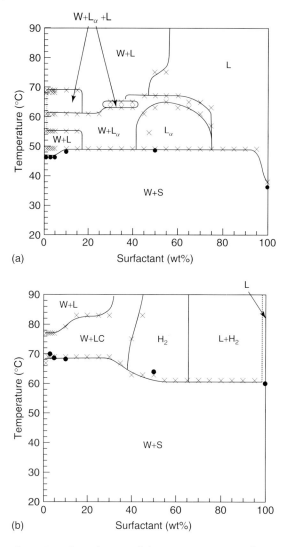

Figure 6.4 Phase diagram of the *pseudobinary water/surfactant systems*. W, aqueous phase; S, solid phase; L_α, lamellar liquid crystalline phase; H_2, inverse hexagonal liquid crystalline phase; and L, isotropic phase. The symbols (×) and (●) represent phase transitions determined by visual observation and by calorimetry, respectively. (a) *Water/nonionic surfactant*. (b) *water/ionic surfactant*. (Reproduced with permission from Ref. [5]).

to 0.2 wt%). The Krafft point of the ionic surfactant at 1 wt% concentration is 67 °C and the melting point of the hydrated nonionic surfactant at 1 wt%, is approximately 48 °C.

In any case if using these surfactants, the emulsification temperature must be higher than 67 °C, which is the Krafft temperature of the ionic surfactant. Below this temperature, solid crystalline phase may coexist with liquid phases, and, therefore, the effective concentration in the liquid is lower than formulated.

In order to know which phases could be present during the formation of paraffin emulsions that are prepared at around 70 °C in industry, the phase behavior of the *pseudoternary water/mixed surfactant system* was studied at high temperature (70 °C). The pseudoternary phase diagram (Figure 6.5) shows two monophasic regions. Compositions with high nonionic/ionic surfactant weight ratio form lamellar liquid crystalline structures, whereas compositions with low nonionic/ionic ratio form inverse hexagonal liquid crystal. The ionic surfactant can be incorporated into the lamellar structure up to a very high nonionic/ionic ratio because of the similar alkyl chain length of both surfactants. They form a very stable lamellar structure. Laughlin and Munyon [40] suggested that in mixed surfactant systems there is an optimum interaction of nonionic and ionic surfactant forming aggregates, which occurs when their alkyl chain lengths are similar (in the present case, the lengths are 21 and 19 carbon atoms for the ionic and the nonionic surfactant, respectively).

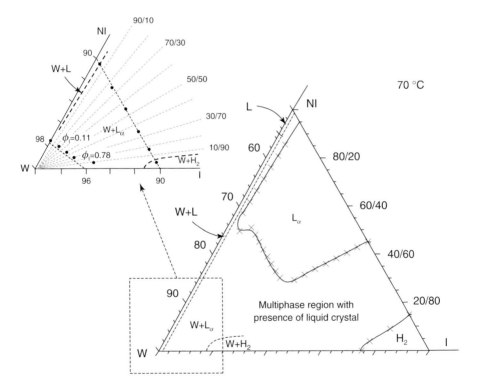

Figure 6.5 Phase diagram of the pseudoternary water/nonionic surfactant/ionic surfactant system at 70 °C. The percentages in this diagram are referred to water content. The diluted region of the diagram is indicated in more detail. ϕ_1 represents the ionic surfactant weight fraction. W, aqueous phase; S, solid phase; H_2, inverse hexagonal liquid crystalline phase; L_α, lamellar liquid crystalline phase; and L, isotropic phase. (Reproduced with permission from Ref. [5].)

The nonionic surfactant requires the presence of the ionic surfactant to form the lamellar structure as no liquid crystal is formed in the binary system, at this temperature. Furthermore, in the multiphase region the coexistence of two liquid crystalline structures was proved. The limits of the different phases found in this region were not well defined.

The phase behavior of the diluted region of the diagram was studied in more detail (down to 98 wt%) at 70 °C (inset in Figure 6.5), as it is the region of interest for emulsions formation. All samples separated into two phases: an aqueous phase and a small volume of a liquid crystalline phase. SAXS results demonstrated that the lamellar phase is present, in the diluted region, even at small nonionic/ionic ratios and it should be pointed out that lamellar liquid crystalline phase has been detected at very low surfactant concentration (down to 0.2 wt%).

Consequently, when paraffin emulsions are prepared (between 1 and 4 wt% of mixed surfactant) aggregates with lamellar structures are present and could absorb onto the particle surface forming a multilayer arrangement that would enhance emulsion stability.

6.3
Formation and Characterization of Paraffin Emulsions

As mentioned in the introduction, preparation of paraffin emulsion is carried out at high temperatures, above the melting points of the paraffin and the nonionic surfactant, and also above the Krafft point of the ionic surfactant. Mújika-Garai et al. [1, 2] and Rodríguez-Valverde et al. [3] studied the influence of preparation parameters on emulsion formation and properties. The surfactant system was the same as that described in the previous section. These authors obtained stable paraffin emulsions with an average particle size of about 1.0 μm and low polydispersity using an ionic/nonionic mixed surfactant system and high-pressure homogenization. At high ionic/nonionic surfactant ratios, a significant change in emulsion properties was observed. The storage modulus (G') increased and particle size and polydispersity decreased, leading to a significant increase in emulsion stability.

In the present case, paraffin emulsions were prepared by high-pressure homogenization at 70 °C, which is above the melting temperatures of the paraffin and the nonionic surfactant and the Krafft temperature of the ionic surfactant. Vilasau et al. [4] studied the influence of the ionic/nonionic surfactant ratio, the total surfactant concentration and the homogenization pressure on the properties of paraffin emulsions. The homogenization pressure was kept constant at 270 atm, and one or two emulsification steps were used.

When emulsifying by a single step, the particle size (Figure 6.6a) decreases slightly with the ionic content, whereas the polydispersity (Figure 6.6b) increases. Figure 6.6 shows the average particle size and the polydispersity as a function of the ionic weight fraction (ϕ_I), for both one and two emulsification steps.

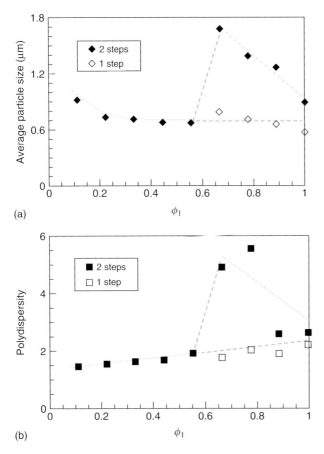

Figure 6.6 (a) Average particle size and (b) polydispersity as a function of the ionic weight fraction (ϕ_I) for one and two emulsification steps. The total surfactant concentration and the homogenization pressure were kept constant. (Reproduced with permission from Ref. [4].)

When applying two emulsification steps, the behavior was more complex. At high ionic content ($\phi_I > 0.56$), emulsions obtained with two steps showed an increase in size and polydispersity, which was clearly attributed to particle aggregation (flocculation), as observed by optical microscopy (Figure 6.7). Therefore, a second emulsification step was detrimental at high ionic content. This behavior was attributed to a low surfactant concentration (2 wt%) that may not be enough to cover larger surface area corresponding to smaller particles, and therefore flocculation occurred. This hypothesis may allow predicting the particle size as a function of surfactant concentration, as described more in detail in Section 6.4.

It is well known that surfactant concentration influences dramatically on emulsion properties. Therefore, this parameter was also studied keeping the ionic

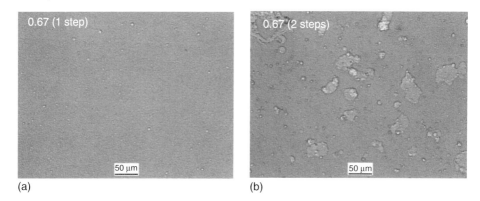

Figure 6.7 Micrographs of the emulsions with an ionic surfactant weight fraction of 0.67 prepared with (a) one and (b) two emulsification steps. (Reproduced with permission from Ref. [4].)

surfactant weight fraction ($\phi_I = 0.33$) and the homogenization pressure (270 atm) constant. The results show that, as expected, the average particle size decreases and polydispersity increases as a function of surfactant concentration (Figure 6.8a). Clearly, higher surfactant concentration leads to larger interfacial areas, and consequently, large particles are obtained at low surfactant concentrations. In addition, both particle size and polydispersity increase abruptly at surfactant concentrations lower than 0.9 wt% (Figure 6.8a, indicated in the graph as a dotted line). At this concentration, flocculation has been observed by optical microscopy, and therefore, the sharp increases in size and polydispersity are artefacts because of such flocculation. This behavior was attributed to an insufficient amount of surfactant molecules to fully cover the surface of the small particles produced by high-pressure homogenization. In these experiments, at 270 atm constant pressure, the smallest size (0.50 μm) was achieved at 4 wt% surfactant. In order to decrease further the particle size, an increase in the homogenization pressure was required.

Another parameter that has an enormous effect on emulsion properties is the homogenization pressure. To study the influence of this parameter, the ionic surfactant weight fraction and the surfactant concentration were kept constant at 0.33 and 4 wt%, respectively (Figure 6.8b). Particle size and polydispersity decrease with pressure up to a critical value (410 atm). Above this pressure, emulsion particles flocculate. This particle aggregation produces again an abrupt increase in size and polydispersity, also indicated as a dotted line (Figure 6.8b). The particle size obtained by high-pressure homogenization (>410 atm) would be too small in relation to the amount of surfactant available for adsorption (4 wt% surfactant). As a consequence, probably there are insufficient surfactant molecules to stabilize an emulsion with such a small particle size (> 0.31 μm, which were obtained at 410 atm). It should be pointed out that emulsions with small size (310 nm) were obtained, which could be considered as nanoemulsions.

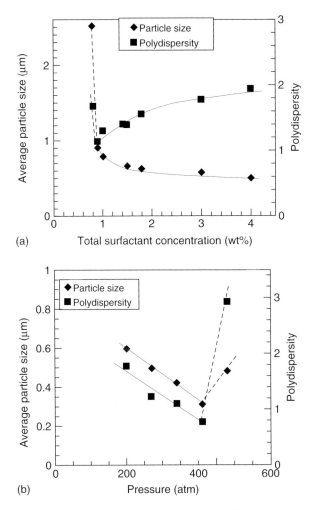

Figure 6.8 Particle size and polydispersity as a function of (a) the total surfactant concentration at constant pressure and (b) the homogenization pressure at constant surfactant concentration. The ionic surfactant weight fraction was kept constant in both cases. The dotted lines remark abrupt changes. (Reproduced with permission from Ref. [4].)

6.4
Control of Particle Size

In industry, reduction of costs can be achieved by optimization of emulsion formulation, minimizing surfactant concentrations while maintaining appropriate emulsion properties that include good stability. Nakajima [41] described an equation to predict the theoretical minimum particle size for nanoemulsion formulations. The desired droplet size (r) could be obtained by choosing an appropriate ratio of oil to surfactant. Nanoemulsions possess higher curvature than conventional emulsions. Therefore, the hydrodynamic layer thickness (monolayer thickness) cannot be neglected. Particle size using *Nakajima's model* could be calculated as follows:

$$r = \left(\frac{3 \times M}{a_s \times N_A \times \rho_o}\right) \times R + \left(\frac{3 \times \alpha \times M}{a_s \times N_A \times \rho_c}\right) + L \tag{6.3}$$

where ρ_o and ρ_c represent the densities of the oil and the surfactant alkyl chain, respectively; α, the weight fraction of the surfactant alkyl chain; R, the oil/surfactant weight ratio; and L, the thickness of the monolayer of hydrated hydrophilic moiety of surfactant.

There are similar equations that also allow predicting the minimum particle size for emulsions, and indeed these equations can be used in the case of paraffin emulsions. Vilasau et al. [4] carried out theoretical calculations considering some well-known assumptions: The surfactant concentration in the aqueous continuous phase corresponds to the critical aggregation concentration (CAC); and the surfactant concentration in the oil phase could be neglected, as the amount of surfactant adsorbed on the interface is much higher. Considering these assumptions the surfactant molecules would be distributed as follows: some molecules are dissolved as monomers in the aqueous continuous phase, at the CAC concentration and almost all molecules are located at the *liquid–liquid* interface. Thus, the total amount of surfactant molecules (S_T) could be defined as the sum of molecules dissolved, as monomers, in the aqueous phase (S_M) and molecules adsorbed at the *liquid–liquid* interface (S_I).

The amount of surfactant dissolved as monomers in the aqueous phase (S_M) could be calculated from the CAC

$$S_M = CAC \times V_T \times (1 - \phi_o) \times M_W \tag{6.4}$$

where V_T represents the total volume of the emulsion, ϕ_o the volume fraction of the oil phase and M_W the surfactant molecular weight.

The surfactant molecules adsorbed at the *liquid–liquid* interface (S_I) could be calculated from the surface/volume ratio of the particles, assuming particles as monodisperse perfect spheres

$$S_I = \frac{3 \times V_T \times \phi_o}{r} \times \frac{M_W}{a_s \times N_A} \tag{6.5}$$

where r represents the radius of the particles, N_A the Avogadro's number, and a_s the surface area per surfactant molecule.

Merging these equations (Eq. (6.4) with Eq. (6.5)), the final expression to obtain the minimum particle size could be expressed as

$$r = \frac{3 \times \phi_o}{a_s \times N_A \times \left(\frac{[S]}{M} + CAC(\phi_o - 1)\right)} \tag{6.6}$$

Then, the minimum particle size could be calculated as a function of surfactant concentration ([S]) by determining the surface area per surfactant molecule (a_s) and the CAC. This equation is denoted as *classical model*.

Nakajima's model (Eq. (6.3)) does not take into account the CAC and the *classical model* (Eq. (6.6)) does not consider the monolayer thickness. Probably, Eq. (6.6) is more appropriate at low surfactant concentration and large particle sizes, where concentrations in the aqueous continuous phase could be closer to CAC. In the present case, the monolayer thickness could be neglected, as it is much thinner in comparison to particle radius.

The Krafft point of the surfactant mixture determined by Mújika-Garai et al. [2] is approximately 48 °C, indicating that the concentration of surfactant molecules as monomers is probably very small at 25 °C. Therefore, a true CMC could not be measured because of this low water solubility. However, the surface tension as a function of surfactant concentration was measured and a sudden change was observed for a specific concentration [4]. This concentration can be considered as the CAC, as probably these aggregates, formed at the solubility limit, are not micelles.

Regarding the surface area per surfactant molecule (a_s), it can be determined by two different methods:

1) **By SAXS measurements**: The surface area that occupies a surfactant molecule in a lamellar liquid crystalline structure could be calculated applying the following equations [42, 43]:

$$a_s = \frac{V_L}{d_L} \tag{6.7}$$

$$d_L = \frac{d \cdot \phi_L}{2} \tag{6.8}$$

where d represents the repeat distance in lamellar phases measured by SAXS, d_L the half thickness of the lipophilic part, ϕ_L the volume fraction of the lipophilic part of the surfactant with respect to the total volume, and V_L the volume of the lipophilic part in a surfactant molecule.

2) **By applying the Gibbs adsorption equation**: The surface area that occupies a surfactant molecule in a micelle could also be calculated from surface tension measurements, using the Gibbs adsorption equation for nonionic surfactants [44] (Eq. (6.9)):

$$\Gamma_1 = \frac{1}{RT} \left(\frac{\partial \gamma}{\partial \ln S} \right)_T \tag{6.9}$$

$$a_s = \frac{10^{16}}{N_A \Gamma_1} \tag{6.10}$$

where Γ represents the surface excess concentration; $\partial \gamma$, the decrease in surface tension; S the surfactant concentration; T, the temperature; and R, the gas constant.

In the present case, there was a discrepancy in the area per molecule depending on the method used. It could be considered that the value obtained by SAXS is more reliable, because of two reasons. Firstly, this surfactant system is composed by a mixture of nonionic and ionic surfactants that is not considered in Eq. (6.9).

Secondly, surface tension data describes the *liquid–gas* interface, in which the area per molecule can be very different form that in *liquid–liquid* interfaces. The advantage of calculations from SAXS data (Eqs. (6.7) and (6.8)) is clear, as these assumptions are not needed. An average area per molecule could be calculated in a mixed surfactant system, which is more representative of the oil–water interface on the emulsion droplets.

Figure 6.9 shows the theoretical models (Eqs. (6.3) and (6.6)) plotted as a function of the total surfactant concentration. In the figure, these simple calculations are compared to experimental data. *Nakajima* and *Classical* equations provided similar results. The possible error in the CAC probably has little influence on the theoretical curves, as CAC $\ll$ ([S]/M). Most likely, the highest error could arise from uncertainty in a_s values that influence more the calculated values.

Interestingly, the theoretical minimum particle size is on agreement with the experimental smallest sizes obtained at different homogenization pressures. Two regions of emulsion formation can be distinguished. All experimental samples are either in the region of stable emulsions or just on the theoretical curves. However, below those curves, formation of stable emulsions is not possible, because the amount of surfactant, available for adsorption, is not enough to fully cover the particles.

The emulsification process was rather efficient as experimental particle sizes were close to the theoretical minimum size. The models assume a low concentration of surfactant in the external phase (equal to CAC), and, therefore, the emulsions are obtained using the minimum amount of surfactant.

Moreover, when preparing emulsions with high-pressure homogenization, small sizes are usually obtained, and therefore, a large interfacial area is generated. Thus, it seems that the particle size depends mostly on surfactant availability for covering particle surface. Consequently, surfactant adsorption probably forms monolayers,

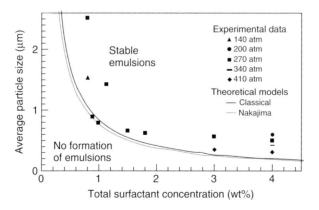

Figure 6.9 Theoretical minimum particle size as a function of surfactant concentration and experimental data for the particle size obtained at different surfactant concentrations and homogenization pressures. (Reproduced with permission from Ref. [4].)

with almost no presence of multilayers (like those illustrated in Figure 6.3), which would require much higher surfactant concentrations.

When emulsification pressure was increased above a critical value, flocculation of particles was observed, and nonstable emulsions were obtained. This aggregation, observed only below the theoretical curves, was due to either insufficient surfactant molecules to fully cover the particles (low surfactant concentration) or too small particles (high interfacial area) resulting from excessive homogenization pressure. By adjusting the total surfactant concentration and/or the homogenization pressure, the minimum particle size could be reached.

Consequently, simple theoretical equations could be a useful tool to formulate stable paraffin emulsions. These results are relevant in industry, allowing the prediction of particle size and optimization of surfactant concentrations. It should be remarked that these results are not only valid for paraffin emulsions, as they can be applied for size prediction of any emulsion, using any oil different from paraffin.

6.5
Stability of Paraffin Emulsions

The fact that at room temperature paraffin is solid like, may greatly decrease instability by either coalescence or Ostwald ripening. Coalescence will be very unlikely because of particle high rigidity, and Ostwald repining may be much reduced, thanks to the low solubility of paraffin in water. However, the systems may be prone to sedimentation and flocculation.

It should be taken into account that emulsion stability depends on the stimuli used to evaluate it, as there is no master test that provides full information about the destabilization mechanisms. Consequently, emulsion stability is usually measured as a function of external stimuli. Three methods have been suggested [6], to assess stability of paraffin emulsions, which simulate destabilization produced in industrial processes: shear produced in a pipeline circuit, freeze–thaw cycles, and addition of electrolytes. Derjaguin, Landau, Verwey and Overbeek (DVLO) theory can be compared to experimental data, in order to evaluate the stabilizing mechanisms involved in the system.

6.5.1
Stability as a Function of Time under Shear (Orthokinetic Stability)

The shear produced in the transportation of an emulsion, from a tank to a container, was simulated pumping the emulsion through a pipeline circuit (Figure 6.10) [6]. Thus, stability of emulsions under shear was studied as a function of time. For this purpose, a model paraffin emulsion was pumped through the pipeline circuit, at room temperature, and samples were analyzed (optical microscopy and laser diffraction) as a function of time. The emulsion was stable during the first 30 min, but particle aggregation was gradually observed at longer times (Figure 6.11).

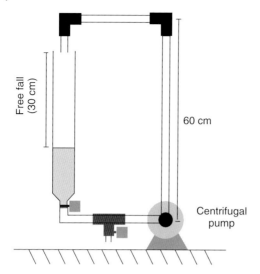

Figure 6.10 Pipeline circuit designed to apply shear to paraffin emulsions. (Reproduced with permission from Ref. [6].)

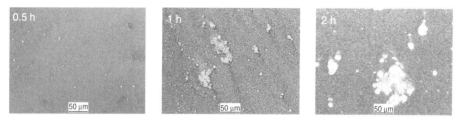

Figure 6.11 Optical micrographs taken after 0.5, 1, and 2 h of applying shear, in the pipeline circuit. (Reproduced with permission from Ref. [6].)

In the present case, the melting point of the paraffin is 52 °C. Therefore, this particle aggregation may not be due to coalescence. Flocculation is the most likely destabilization mechanism, as confirmed by optical microscopy. It was attributed to the high shear produced inside the centrifugal pump and in the free fall.

High shear leads to flocculation when the kinetic energy of particles overcomes the repulsive energy barrier. This phenomenon is denominated orthokinetic flocculation [45]. After 2 h, the high degree of flocculation produced a plugging in the pipeline circuit, after a high increase in viscosity. Therefore, orthokinetic flocculation can be very fast, if the repulsive energy barrier is not high.

6.5.2
Stability as a Function of Freeze–Thaw Cycles

Phase separation can also be accelerated by applying cycles of ultracentrifugation and thermal treatment that are usually denominated freeze–thaw cycles. These stability tests can produce separation of emulsions in three different layers, which are clearly identified above the melting temperature of the paraffin: (i) an oil phase, (ii) a concentrated emulsion (remaining residual emulsion layer), and (iii) a third fraction of very diluted small paraffin particles dispersed in an aqueous phase (water-rich layer). The volume percentage of the residual emulsion was measured at the end of each cycle and photographs were taken at 70 °C (all phases were in a liquid state). The stability was quantified as the volume of the residual emulsion layer that was larger for more stable systems.

Stability was determined as a function of ionic surfactant weight fraction, while keeping constant all other parameters. Figure 6.12a shows the residual emulsion volume plotted as a function of the ionic weight fraction. The results indicate that stability increases gradually with the ionic surfactant concentration. The higher the ionic surfactant content, the more stable is the emulsion. Figure 6.12b illustrates this trend in a photograph, taken after five freeze–thaw cycles. In this example, the ionic surfactant stabilizes much more effectively than the nonionic surfactant, indicating probably a predominantly electrostatic stabilization.

The total surfactant concentration was also studied to determine its influence on emulsion stability. It was varied from 0.8 to 4 wt% and the volume

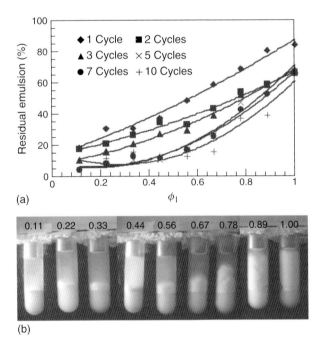

Figure 6.12 Evaluation of stability against freeze–thaw cycles for emulsions as a function of the ionic surfactant weight fraction (ϕ_I). (a) Volume percentage of residual emulsion as a function of the ionic weight fraction, for different cycles and (b) visual aspect of the emulsions after five cycles. The numbers indicate the ionic weight fraction values. (Reproduced with permission from Ref. [6].)

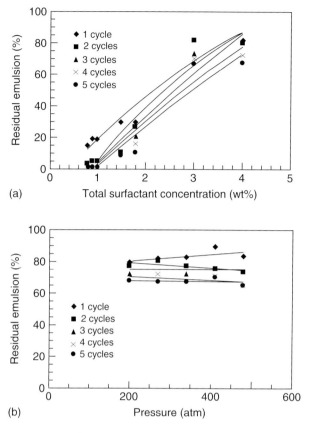

Figure 6.13 Volume percentage of residual emulsion as a function of (a) the total surfactant concentration, keeping constant homogenization pressure and (b) the homogenization pressure, keeping constant total surfactant concentration. The ionic surfactant weight fraction was kept constant in both cases. (Reproduced with permission from Ref. [6].)

of residual emulsion was measured as a function of surfactant concentration (Figure 6.13a). The higher the surfactant content, the higher is the stability against freeze–thaw cycles. This is not surprising, as increasing surfactant concentration allows decreasing particle size, because larger interfacial area can be covered by surfactant adsorption. It can be observed that the most stable emulsion, prepared at 4 wt% surfactant concentration, possesses the smallest particle size ($\approx$0.5 µm).

Stability was also determined as a function of homogenization pressure, varied from 140 to 540 atm (Figure 6.13b). All samples were very stable and only a small upper layer of supernatant (oil phase) was observed. It should be pointed out that these samples were prepared with a relatively high surfactant concentration (4 wt%) that allowed small sizes between 0.3 and 0.6 µm, resulting in high stability. Emulsions prepared at pressure either lower than 200 or higher than 480 atm were not stable, and consequently freeze–thaw cycles could not be carried out. In the former ($P < 200$ atm), particle size was large and spontaneous creaming was observed. In the latter ($P > 480$ atm), particles appeared already flocculated after homogenization.

The results of freeze–thaw cycles are probably related to the fact that large particles cream faster, whereas the smallest particles remain stable in the aqueous continuous phase. During the thermal treatment, probably the largest particles coalesce, and therefore, the stability against freeze–thaw cycles is increased when particle size is decreased. In conclusion, these results illustrate that the homogenization pressure and the surfactant concentration require careful adjustments, in order to reach the minimum size (Figure 6.9) and the highest stability. Electrostatic repulsions could also play an important role, and, therefore, the influence of electrolytes on emulsion stability should be systematically studied.

6.5.3
Stability as a Function of Electrolytes

In industrial applications, paraffin emulsions are diluted generally using water containing high electrolyte concentration. Therefore, emulsion stability was studied as a function of electrolyte concentration by means of critical coagulation concentration (CCC) determinations [6]. Two electrolytes were used: sodium chloride (NaCl) because it is a 1 : 1 electrolyte, which often satisfies DLVO theory; and calcium chloride ($CaCl_2$), selected as a 2 : 1 electrolyte. Figure 6.14 shows the results of the CCC of NaCl (CCC_{NaCl}) for an emulsion stabilized solely with the ionic surfactant ($\phi_I = 1$). Figure 6.14a shows diluted emulsions after 24 h of electrolyte addition. In this case, samples with low electrolyte concentration ([NaCl] < 20 mM) are very turbid, indicating that paraffin particles are well dispersed in the continuous phase. However, samples with higher electrolyte concentrations (20 mM < [NaCl] < 100 mM) show a creamed upper layer, produced because of particle flocculation. The lower layer (mainly aqueous solution) possesses low turbidity, as a smaller amount of particles is remaining in suspension. Samples with higher electrolyte concentration ([NaCl] $\geq$ 100 mM) show a creamed layer and a highly transparent aqueous dispersion. Therefore, these samples are considered to be completely coagulated.

Optical microscopy was used to confirm whether samples were partially flocculated or not. Figure 6.14b shows micrographs of samples with 15 and 17 mM NaCl. Flocculation was observed at 17 mM, while it appeared mainly nonflocculated at 15 mM (Figure 6.14b). Therefore, emulsions may be highly dependent on small variation on electrolyte concentrations.

Electrophoretic mobility measurements are usually performed to evaluate the electrostatic stability. The results [4] have shown that electrophoretic mobility increases with the ionic weight fraction, before reaching a plateau, probably because of saturation of charges at the oil–water interface. The zeta potential was calculated using the Smoluchowski equation following the Stern model [46], and the maximum value at the saturation point was -75 ± 3 mV [4]. This zeta potential

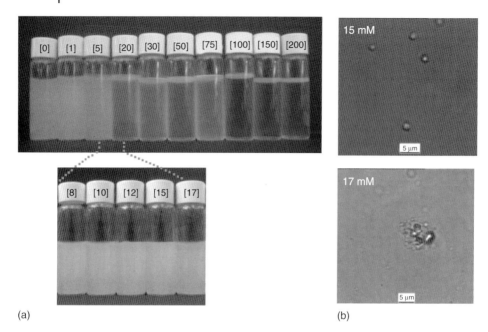

Figure 6.14 Determination of the critical coagulation concentration of NaCl (CCC_{NaCl}) for the emulsion with a $\phi_I = 1$. (a) Diluted emulsions with a NaCl concentration from 0 to 200 mM and 8 to 17 mM. (b) Optical micrographs of the diluted emulsion with [NaCl] = 15 mM and [NaCl] = 17 mM. (Reproduced with permission from Ref. [6].)

is rather high, indicating that the particles are highly charged, as often found in other systems stabilized with mixtures of ionic and nonionic surfactants [47, 48].

Theoretical simulations were performed using DLVO theory for colloidal stability [49, 50], in order to explain the experimental data of the CCC determinations. DLVO theory does not consider steric repulsions and other structural interactions. However, the determination of a total interaction potential is very useful to predict whether an emulsion is stable, flocculated, or strongly coagulated. Experimental values of CCC, for NaCl, were compared to DLVO predictions [6]. NaCl was used because this 1 : 1 electrolyte usually fits into DLVO theory. The emulsion stabilized solely with the ionic surfactant ($\phi_I = 1$) was also used for the simulation, and, therefore, steric repulsions were neglected. DLVO theory takes into account van der Waals attractive forces and repulsive forces generated from the electrical double layer of the particles. The van der Waals potential can be calculated using the following equation for two spherical particles [51],

$$W_{VDW} = \frac{-Ar}{12D} \tag{6.11}$$

where A represents the Hamaker constant; r, the particle radius; and D, the distance between particles. It is known that the Hamaker constant is not truly constant, and may depend on electrolyte concentration, as indicated in Eq. (6.12) [52],

$$A = A_{\nu=0}(2\kappa D)e^{-2\kappa D} + A_{\nu > \nu_1} \tag{6.12}$$

where $A_{\upsilon=0}$ and $A_{\upsilon>0}$ are zero and nonzero frequency contributions of the Hamaker constant, respectively, and κ the inverse of the Debye length.

The Hamaker constant, for the paraffin–water–paraffin symmetric system, was considered the same as in the hexadecane–water–hexadecane system, 5.0×10^{-21} J [53]. At low electrolyte concentrations, this value could be considered constant, despite the fact that it slightly depends on electrolyte concentration.

The electrostatic potential due to the electrical double layer of the particles can be determined from the Derjaguin approximation [49] and considering a weak overlap, the equation takes the following form:

$$W_E = \left(\frac{64\pi kTr[\text{NaCl}] \tanh^2(ze\psi_0/4kT)}{\kappa^2} \right) e^{-\kappa D} \tag{6.13}$$

where k represents the Boltzmann's constant; T, the absolute temperature; [NaCl], the electrolyte concentration; z, the valence of the electrolyte; e, the electron charge; and ψ_0 the surface potential of the particle. Such an approximation is only valid for the interactions between two similar surfaces at constant potential [51]. The inverse of the Debye length (κ) is known as the *Debye–Hückel parameter* and it can be calculated as follows:

$$\kappa = \sqrt{\frac{2z^2 e^2 [\text{NaCl}]}{\varepsilon_r \varepsilon_0 kT}} \tag{6.14}$$

where ε_r represents the relative dielectric constant of water and ε_0 the vacuum permittivity.

Regarding the surface potential (ψ_0), it could be calculated by two different methods:

1) **Using Grahame's equation**: It could be assumed that all the charged surfactant molecules are adsorbed on the particle surface and the concentration of ionic surfactant molecules in the aqueous media is close to zero. In this case, surface charge density could be calculated from known parameters, such as surfactant concentration and particle radius, according to

$$\sigma = \frac{[S]N_A er}{V_T 3\phi_0} \tag{6.15}$$

where $[S]$ represents the surfactant concentration; N_A, the Avogadro's number; e, the electron charge; r, the particle radius; V_T, the total volume; and ϕ_0, the oil phase volume fraction.

The surface charge density and the surface potential are related by the already mentioned Grahame's equation [51]:

$$\sigma = \sqrt{8\varepsilon_r \varepsilon_0 kT[\text{NaCl}]} \sinh\left(\frac{e\psi_0}{2kT}\right) \tag{6.16}$$

2) **Using Debye–Hückel equation**: The surface potential can be calculated by determining zeta potential and considering the exponential decay of electrostatic potential [50].

$$\psi_0 = \zeta e^{\kappa x} \tag{6.17}$$

where ζ represents the zeta potential; κ, the inverse of the Debye length; and x, the hydrodynamic layer thickness. In this case, uncertainty could come from Smoluchowski approximation (zeta potential) and/or from hydrodynamic layer thickness. An approximate value for x is assumed, considering the dimension of a single water molecule ($x = 0.25$ nm).

It should be pointed out that calculations of the surface potential by the Grahame's equation are based on complete adsorption of the ionic surfactant, which is highly unlikely. Therefore, in many cases, the use of the Debye–Hückel equation to calculate the surface potential seems to be more precise, if good measurements of zeta potential can be obtained.

For this reason, determinations of electrophoretic mobility, as a function of electrolyte concentration, are required. Zeta potential can be calculated from the experimental values of electrophoretic mobility, using the simple Smoluchowski equation. As an illustration, the results of zeta potential of a paraffin emulsion stabilized solely with an ionic surfactant are shown in Figure 6.15 [6]. High absolute values were obtained, indicating that electrostatic repulsion is the main stabilizing mechanism. Initially, the absolute value of zeta potential increased up to 90 mV, at 10 mM NaCl. However, increasing further the electrolyte concentration produces a final decrease, down to approximately 20 mV, at 500 mM NaCl. This behavior was expected and consistent to that already described in the literature [54–56]. At very low electrolyte concentration, the adsorbed chains could be extended into the aqueous phase, and the shear plane (on which zeta potential is defined) would be far away from the surface. Increasing slightly [NaCl], the shear plane could move toward the surface and consequently, zeta potential increases. However, by increasing [NaCl] further, the electrostatic repulsion layer also shrinks, because of charge screening induced by electrolyte. This behavior usually produces a decrease in absolute values of zeta potential at high-electrolyte concentrations.

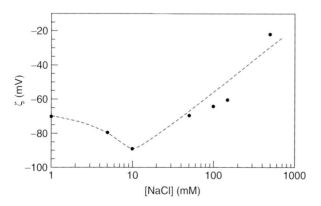

Figure 6.15 Zeta potential as a function of NaCl concentration, calculated by applying Smoluchowski equation to experimental electrophoretic mobility. (Reproduced with permission from Ref. [6].)

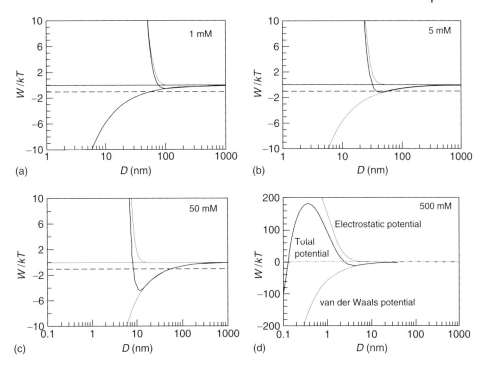

Figure 6.16 Interaction potential as a function of the distance between two particles of the emulsion stabilized solely with ionic surfactant, for different electrolyte concentrations. (a) [NaCl] = 1 mM; (b) [NaCl] = 5 mM; (c) [NaCl] = 50 mM; and (d) [NaCl] = 500 mM. (Reproduced with permission from Ref. [6].)

The DLVO equations mentioned above were used to calculate the total interaction potential for different electrolyte concentrations as a function of the distance between two particles (Figure 6.16) [6]. As mentioned before, the total potential was calculated as the sum of the electrostatic repulsion and the van der Waals attraction, neglecting steric repulsion. The emulsion stabilized solely with the ionic surfactant was used for the simulation. The repulsive contribution (electrostatic potential) and the attractive contribution (van der Waals potential) were also plotted. The potential is expressed as kT, where k is the Boltzmann constant and T the absolute temperature. The minimum energy for a stable colloidal state is considered to be $-1kT$, as Brownian motion ($\approx kT$) cannot compensate a deeper energy minimum.

In the present case, the electrostatic repulsive forces predominate over weak van der Waals attractive forces, at 1 mM NaCl concentration (Figure 6.16a), and the particle dispersion remains stable. If increasing the NaCl concentration to 5 mM, the electrostatic repulsion still predominates over the repulsive forces and coagulation would be prevented (Figure 6.16b). The minimum in the energy profile is only approximately $-1kT$, and therefore, it would produce a rather weak reversible flocculation. The electrostatic repulsion decreases when increasing [NaCl] to 50 mM (Figure 6.16c). At short separation distances, there is still a high-energy barrier, which would prevent coagulation. However, at long separation distances, the van der Waals attractive forces are higher than the electrostatics and the particles may flocculate. When [NaCl] was very high (500 mM), the energy profile shows a primary minimum (coagulation at short distances) and a secondary minimum (flocculation at longer distances), with a much lower energy barrier (Figure 6.16d). This indicates that the van der Waals attractive forces predominate over weak electrostatic repulsion and coagulation would occur. However, this coagulation may be slow, since the remaining energy barrier is still high, at approximately 180 kT.

As mentioned before, experimental data indicates that weak flocculation occurred at approximately 17 mM NaCl, and strong coagulation above 100 mM NaCl. Consequently, the DLVO calculation can explain the overall tendencies observed by experiment. However, coagulation seems to occur earlier than expected by theory. Consequently, from the point of view of coagulation, the system is less stable than that predicted by theory. This discrepancy could be attributed to the presence of electrolytes other than NaCl. As mentioned previously, paraffin emulsions are formulated with an excess of base, in order to achieve complete neutralization of the fatty acid. Therefore, it may result in a rather large excess of counterion, resulting in higher ionic strength than expected, and in a lower stability.

In the present work, the stability against electrolyte concentration (CCC) was also studied using a 2 : 1 electrolyte. According to the Schultz–Hardy rule, multivalent cations such as Ca^{2+} are much more effective in inducing flocculation [57]. CCC_{CaCl_2} has been determined as a function of the ionic/nonionic ratio. The results are indicated in Figure 6.17, showing that emulsions prepared without ionic surfactant ($\phi_I = 0$) already appear flocculated after emulsification, considering their CCC as zero. Addition of ionic surfactant produces an increase in CCC_{CaCl_2}, as expected. However, it reaches a maximum value, at intermediate ϕ_I and decreases down to 4.0 mM at $\phi_I = 1$.

A naïve assumption would be that surface charge density should be maximum at $\phi_I = 1$. However, Ca^{2+} ions probably adsorb on particle surface by forming

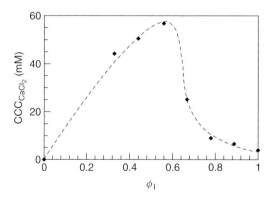

Figure 6.17 Determination of the critical coagulation concentration of $CaCl_2$ (CCC_{CaCl_2}) as a function of the ionic surfactant weight fraction (ϕ_I). (Reproduced with permission from Ref. [6].)

complexes with carboxylates, and reducing effective charges around the particles. Therefore, this may explain that CCC decreases dramatically at high ionic content. Consequently, a combination of steric and electrostatic forces is essential for stabilizing against Ca^{2+}. Electrostatic charges are probably very important in stabilization, but they can be screened out by Ca^{2+} ions, and addition of a nonionic surfactant is required to maximize the CCC.

6.6 Conclusions

In this chapter a systematic work on surfactant selection, formation and properties of paraffin emulsions is presented. The main objective has been to study emulsion formulation, controlling its particle size and understanding stability mechanisms. The results, referred to paraffin emulsions, may be also applied to other emulsions, when using oils different from paraffin.

Traditionally, paraffin emulsions were stabilized with ionic surfactant providing electrostatic stability. However, the addition of a nonionic surfactant to the system results in enhanced electrosteric stabilization, using surfactants with long alkyl chains, which may adsorb strongly on lipophilic particle surface. Moreover, stability can also be improved by the presence of solid-like rigid surfactant monolayers, which are formed cooling down below the Krafft point of the ionic surfactant and the melting temperature of the nonionic surfactant. Therefore, phase behavior has to be considered when selecting the most appropriate conditions for emulsification. Furthermore, it should be taken into account that liquid crystalline phases may be present in some systems during emulsification, at very low surfactant concentrations.

The minimum size of emulsion droplets can be predicted by using simple equations, based on surfactant coverage of interfacial area, with a good agreement between theory and experiment. These equations also constitute a useful tool to predict stability of emulsions prepared by high-energy emulsification methods, and they can be relevant in industrial applications.

Stability evaluation depends on the method used to study it. For this reason, stability should be measured as a function of different stimuli. For example, some emulsions may present moderate stability under shear, high stability under freeze–thaw cycles, and low stability under addition of electrolyte. The most stable emulsions against freeze–thaw cycles are those with high ionic/nonionic surfactant ratio, as electrostatic repulsion plays a very important role. Moreover, stability increases at high-surfactant concentration and high-homogenization pressure that produce emulsions with smaller sizes. Emulsions may present low stability against high-electrolyte concentrations because of the screening of the charges at the particle surface. However, this stability can be increased by the addition of long-chain nonionic surfactants that enhance steric repulsion.

The stability of emulsions, formulated solely with ionic surfactant, is consistent with DLVO theory, considering only the electrostatic repulsion and neglecting the steric interactions. In the presence of Ca^{2+}, the stability is much lower than with NaCl, as expected, and depends mostly on the balance between electrostatic and steric repulsions. Nonionic surfactant is required to maximize CCC, probably becauses of adsorption of Ca^{2+} ions at particle surface.

Acknowledgments

The authors greatly acknowledge the Spanish Ministry of Science and Innovation (PET2006-0582 and CTQ2008-06892-C03-01 projects), Generalitat de Catalunya (SGR-00961), and REPSOL-YPF for the financial support. The authors also acknowledge C. Solans (IQAC-CSIC) and R. Mújika (REPSOL-YPF) for useful discussions.

References

1. Mújika-Garai, R., Covián-Sánchez, I., Tejera-García, R., Rodríguez-Valverde, M.A., Cabrerizo-Vílchez, M.A., and Hidalgo-Álvarez, R. (2005) *J. Disper. Sci. Technol.*, **26**, 9–18.
2. Mújika-Garai, R., Aguilar-García, C., Juárez-Arroyo, F., Covián-Sánchez, I., Nolla, J., Esquena, J., Solans, C., Rodríguez-Valverde, M.A., Tejera-García, R., Cabrerizo-Vílchez, M.A., and Hidalgo-Álvarez, R. (2007) *J. Disper. Sci. Technol.*, **28**, 829–836.
3. Rodríguez-Valverde, M.A., Tejera-García, R., Cabrerizo-Vílchez, M.A., Hidalgo-Álvarez, R., Nolla-Anguera, J., Esquena-Moret, J., Solans-Marsa, C., Mújika-Garai, R., Aguilar-García, C., Arroyo, F.J., and Covián-Sánchez, I. (2006) *J. Disper. Sci. Technol.*, **27**, 155–163.
4. Vilasau, J., Solans, C., Gómez, M.J., Dabrio, J., Mújika-Garai, R., and Esquena, J. (2011) *Colloids Surf. A: Physicochem. Eng. Aspects*, **392**, 38-44.
5. Vilasau, J., Solans, C., Gómez, M.J., Dabrio, J., Mújika-Garai, R., and Esquena, J. (2011) *Colloids Surf. A: Physicochem. Eng. Aspects*, **384**, 473–481.
6. Vilasau, J., Solans, C., Gómez, M.J., Dabrio, J., Mújika-Garai, R., and Esquena, J. (2011) *Colloids Surf. A: Physicochem. Eng. Aspects*, **389**, 222-229.
7. Balducci, F., Harper, C., Meinlschmidt, P., Dix, B., and Sanasi, A. (2008) *Drvna Industrija*, **59**, 131–136.
8. Batalla, L., Nuñez, A.J., and Marcovich, N.E. (2005) *J. Appl. Polym. Sci.*, **97**, 916–923.
9. Goswami, T., Kalita, D., Ghosh, S.K., and Rao, P.G. (2006) *IPPTA: Q. J. Indian Pulp Paper Tech. Assoc.*, **18**, 39–43.
10. Iskanderani, F.I. (2008) *Int. J. Polym. Mater.*, **57**, 979–995.
11. Khedari, J., Charoenvai, S., and Hirunlabh, J. (2003) *Build. Environ.*, **38**, 435–441.
12. Nemli, G., Kirci, H., Serdar, B., and Ay, N. (2003) *Ind. Crops Prod.*, **17**, 39–46.
13. Ntalos, G.A. and Grigoriou, A.H. (2002) *Ind. Crops Prod.*, **16**, 59–68.
14. Papadopoulos, A.N. and Hague, J.R.B. (2003) *Ind. Crops Prod.*, **17**, 143–147.
15. Xu, X., Yao, F., Wu, Q., and Zhou, D. (2009) *Ind. Crops Prod.*, **29**, 80–85.
16. Dorset, D.L. (1997) *J. Phys. D: Appl. Phys.*, **30**, 451–457.
17. Meyer, G. (2006) *Erdoel Erdgas Kohle*, **122**, 16–18.
18. Musser, B.J. and Kilpatrick, P.K. (1998) *Energy Fuels*, **12**, 715–725.
19. Chapman, D. (1965) *The Structure of Lipids by Spectroscopic and X-ray Techniques*, John Wiley & Sons, Inc., New York.
20. von Berlepsch, H., Hofmann, D., and Ganster, J. (1995) *Langmuir*, **11**, 3676–3684.
21. Pressl, K., Jørgensen, K., and Laggner, P. (1997) *Biochim. Biophys. Acta*, **1325**, 1–7.

22. Freund, M., Csikós, R., Keszthelyi, S., and Mózes, G.Y. (1982) *Paraffin Products Properties, Technologies, Applications*, Elsevier, New York.
23. Amthor, J. (1972) *Holz als Roh- und Werkstoff*, **30**, 422–429.
24. Walstra, P. (1983) *Formation of Emulsions in Encyclopaedia of Emulsion Technology*, Marcel Dekker, New York.
25. Ali, A.A. and Mulley, B.A. (1978) *J. Pharm. Pharmacol.*, **30**, 205–213.
26. Eccleston, G.M. (1990) *J. Soc. Cosmet. Chem.*, **41**, 1–22.
27. Friberg, S. (1971) *J. Colloid Interface Sci.*, **37**, 291–295.
28. Friberg, S., Mandell, L., and Larsson, M. (1969) *J. Colloid Interface Sci.*, **29**, 155–156.
29. Friberg, S. and Solyom, P. (1970) *Polymere*, **236**, 173–174.
30. Friberg, S.E. and Solans, C. (1986) *Langmuir*, **2**, 121–126.
31. dos Santos, O.D.H., Miotto, J.V., de Morais, J.M., da Rocha-Filho, P.A., and de Oliveira, W.P. (2005) *J. Disper. Sci. Technol.*, **26**, 243–249.
32. Morais, G.G., Santos, O.D.H., Oliveira, W.P., and Rocha Filho, P.A. (2008) *J. Disper. Sci. Technol.*, **29**, 297–306.
33. Pasquali, R.C. and Bregni, C. (2006) *Ars Pharm.*, **47**, 219–237.
34. Tadros, T. and Vincent, B.P. (1983) *Formation of Emulsions in Encyclopaedia of Emulsion Technology*, Marcel Dekker, New York.
35. Aveyard, R., Binks, B.P., Esquena, J., Fletcher, P.D.I., Bault, P., and Villa, P. (2002) *Langmuir*, **18**, 3487–3494.
36. Aveyard, R., Binks, B.P., Esquena, J., Fletcher, P.D.L., Buscall, R., and Davies, S. (1999) *Langmuir*, **15**, 970–980.
37. Friberg, S. and Mandell, L. (1970) *J. Am. Oil Chem. Soc.*, **47**, 149–152.
38. Adam, C.D., Durrant, J.A., Lowry, M.R., and Tiddy, G.J.T. (1984) *J. Chem. Soc., Faraday Trans. 1*, **80**, 789–801.
39. Lide, D.R. (2004) *Handbook of Chemistry and Physics (CRC)*, 86th edn, CRC Press, Boca Ratón, FL, London, New York, Washington, DC.
40. Laughlin, R.G. and Munyon, R.L. (1984) *Chem. Phys. Lipids*, **35**, 133–142.
41. Nakajima, H. (1997) *Microemulsions in Cosmetics in Industrial Applications of Macroemulsiones*, Marcel Dekker, New York.
42. Kunieda, H., Kabir, H., Aramaki, K., and Shigeta, K. (2001) *J. Mol. Liq.*, **90**, 157–166.
43. Sagitani, H. and Friberg, S.E. (1983) *Colloid Polym. Sci.*, **261**, 862–867.
44. Gibbs, J.W. (1928) *The Collected Works of J. W. Gibbs*, Longmans, Green, London.
45. Smoluchowksi, M. (1917) *Z. Phys. Chem.*, **92**, 129–168.
46. Roland, I., Piel, G., Delattre, L., and Evrard, B. (2003) *Int. J. Pharm.*, **263**, 85–94.
47. Backfolk, K., Olofsson, G., Rosenholm, J.B., and Eklund, D. (2006) *Colloids Surf., A: Physicochem. Eng. Aspects*, **276**, 78–86.
48. Goloub, T.P. and Pugh, R.J. (2005) *J. Colloid Interface Sci.*, **291**, 256–262.
49. Derjaguin, B. (1934) *Kolloid-Zeitschrift*, **69**, 155–164.
50. Verwey, E.J.W., Overbeek, J.T.G., and Nes, Kv. (1948) *Theory of the Stability of Lyophobic Colloids; The Interaction of Sol Particles Having An Electric Double Layer*, Elsevier Publishing Company, New York.
51. Israelachvili, J. (1992) *Intermolecular & Surface Forces*, Academic Press Limited, London.
52. Mahanty, J. and Ninham, B.W. (1976) *Dispersion Forces*, Academic Press, London, New York.
53. Hough, D.B. and White, L.R. (1980) *Adv. Colloid Interface Sci.*, **14**, 3–41.
54. Jódar-Reyes, A.B., Ortega-Vinuesa, J.L., and Martín-Rodríguez, A. (2006) *J. Colloid Interface Sci.*, **297**, 170–181.
55. Meijer, A.E.J., van Megen, W.J., and Lyklema, J. (1978) *J. Colloid Interface Sci.*, **66**, 99–104.
56. Van Der Put, A.G. and Bijsterbosch, B.H. (1983) *J. Colloid Interface Sci.*, **92**, 499–507.
57. Israelachvili, J.N. (1991) *Intermolecular and Surface Forces*, 2nd edn, Academic Press, London, San Diego, CA.

7
Polymeric O/W Nano-emulsions Obtained by the Phase Inversion Composition (PIC) Method for Biomedical Nanoparticle Preparation

Gabriela Calderó and Conxita Solans

7.1
Introduction

Nano-emulsions are a kind of emulsions having droplet diameters below 1 μm, although from a more general point of view, only those with droplet sizes up to 500 nm are considered of interest for most applications [1]. Nano-emulsions show appealing properties, especially in the biomedical field, for the solubilization of drugs and also for the preparation of therapeutic or diagnostic nanoparticle dispersions [2, 3]. When formulating pharmaceutical products, stability, safety, and efficacy are of utmost importance. Concerning stability, because of their small size, nano-emulsions are stable against sedimentation and creaming. In relation to safety, biocompatible components that are appropriate for the route of administration should be chosen. The drug to be incorporated into the template nano-emulsion may also have specific requirements concerning its solubility and stability. The solubility of the drug in the dispersed phase should be high enough to ensure that the therapeutic concentration at a reasonable dosage is achieved. Drugs that may suffer hydrolysis when in contact with water are good candidates to be solubilized in the dispersed phase of oil-in-water (O/W) systems and/or encapsulated into nanoparticles. In addition, if thermolabile components are present, preparation procedures that can be performed in the temperature range at which the drug is stable are mandatory. This chapter focuses on the use of nano-emulsions for the preparation of polymeric nanoparticles. Nanoparticles for medical applications have been defined as solid colloidal particles with a size ranging between 1 and 1000 nm [4, 5], regardless of their preparation method, chemical nature, or structure. Some of their biomedical uses include the delivery of drugs to subcellular compartments or across biological barriers, enhancement of drug bioavailability, passive tumor targeting, thermal tumor ablation, or improved imaging among others.

Emulsion Formation and Stability, First Edition. Edited by Tharwat F. Tadros.
© 2013 Wiley-VCH Verlag GmbH & Co. KGaA. Published 2013 by Wiley-VCH Verlag GmbH & Co. KGaA.

7.2
Phase Inversion Emulsification Methods

Nano-emulsions can be prepared by dispersion or high-energy methods, implying external energy provided, generally, either by rotor/stator devices (e.g., ultraturrax) [6], ultrasound generators [6–8], or high-pressure homogenizers [9]. In the last decades, much attention has been given to condensation or low-energy methods that make use of the chemical energy stored in the system. Self-emulsification methods [10–12] and phase inversion methods are the most well known. In phase inversion methods, emulsification is triggered by phase transitions that produce changes in the spontaneous curvature of the surfactant film. These transitions often involve the inversion of the surfactant film curvature from positive (O/W) to negative (water-in-oil, W/O) or vice versa. However, uniform and nanosized emulsions can be obtained from structures having a surfactant film with an average zero curvature, which may experience phase transitions to structures with positive or negative curvature. This occurs, for example, when starting emulsification from bicontinuous microemulsions [13, 14], bicontinuous cubic phases [15], or lamellar liquid crystalline phases [16]. Phase transitions may be triggered by temperature changes in water/nonionic ethoxylated surfactant/oil systems owing to the change in the affinity of ethoxylated surfactants from water to oil at increasing temperature [17]. This approach is called phase inversion temperature (PIT) method [13, 18, 19]. For nonethoxylated surfactants and in many of its applications, however, where it is not possible to modify temperature over a broad span because of stability issues or energy and time savings, achieving phase inversion by changing composition at constant temperature, phase inversion composition (PIC) method is preferred [3, 20, 21]. In this approach, the continuous phase is added to the dispersed phase containing the surfactant or surfactant mixture, until phase inversion occurs. The PIC method has been applied for the preparation of O/W [3, 21–23] and W/O [24–26] nano-emulsions in systems containing either nonionic ethoxylated surfactants [3, 21, 24, 25] or mixtures thereof with ionic surfactants [27, 28]. In mixed nonionic/ionic surfactant systems, phase transitions can be obtained by diluting in brine [29], by varying the ionization degree of the surfactant molecule [15], and so on. Average droplet size and polydispersity of the nano-emulsions determine many of their properties, such as their optical appearance, stability, etc. The droplet size of nano-emulsions prepared by the PIC method has been shown to depend on a series of composition and process parameters, such as O/S ratio [3, 21, 30, 31] (G. Calderó *et al.*, unpublished results), hydrophilic–lipophilic balance (HLB) of the surfactant mixture [30], dispersed phase fraction, temperature [30], addition time [21, 23, 32], mixing rate [23]... In O/W nano-emulsions, one of the most influencing parameters is the O/S ratio. The expected tendency is an increase, generally linear, of the droplet size at increasing O/S ratio [3, 21, 30]. However, several parameters can play a role at a time, which would explain that sometimes droplet size varies according to a quadratic equation at increasing O/S ratios [30] (G. Calderó *et al.*, unpublished results). Other composition variables influencing the droplet size and hence the stability of the nano-emulsions are the mutual solubilities of the two phases and the amount and types of surfactants [7].

As nano-emulsions are nonequilibrium systems, their properties depend also on the preparation method. Parameters such as the emulsification path [16, 20] or the speed of addition of the components [33] determine the features of the emulsion. In addition, the preparation of larger amount of nano-emulsions is a matter of concern, although it has been shown that the scale up of nano-emulsions prepared by the PIC and PIT methods can be performed [23].

7.3
Aspects on the Choice of the Components

The preparation method of nanoparticles is strongly conditioned by their nature, while its choice depends on the specific requirements of the aimed application. In this context, polymeric nanoparticles can be prepared in a simple way from preformed polymers by the nano-emulsion approach. This method does not require sophisticated equipment, especially when low-energy methods are employed. Although nano-emulsions can be used as nanoreactors for monomer polymerization (also known as *miniemulsion polymerization*) in the dispersed droplets, this way of preparation shows disadvantages mainly because of the need to use reactants that may be potentially toxic and require purification steps to remove their excess. In addition, other components of the nano-emulsion, such as the surfactant or actives, may get involved in unwanted reactions. Therefore, the use of preformed pharmaceutically accepted polymers is an attracting alternative. The most well-known method of obtaining nanoparticles from O/W nano-emulsions is the solvent evaporation method. For this purpose, the polymer should be insoluble in water, and it is dissolved in the oil component prior to nano-emulsion preparation. Typical preformed polymers are polyesters (polylactic acid, poly(lactic-*co*-glycolic) acid, poly(ε-caprolactone)), poly(methyl methacrylates), cellulose derivatives (cellulose acetate butyrate, ethylcellulose, cellulose acetate phtalate), and so on. In addition to the nature of the polymer, its molecular weight plays an important role in the performance *in vivo* of the final nanoparticle, affecting its blood clearance and its biodistribution. For example, it has been reported that the renal elimination of certain polymers is granted if molecular weight is below 30 000 g mol^{-1} [34, 35]. This threshold is, especially, important when nonbiodegradable polymers are administered by the parenteral route, to avoid accumulation [36]. At a cellular level, molecular weight is also important. Thus, for gene therapy purposes, penetration through the nuclear membrane is possible if the polymer chain is linear and its molecular weight is below 22 000 g mol^{-1} [37], although a higher limit of 50 kDa has also been reported [38]. Suitable solvents used as oil component are those immiscible or only partially miscible with water, in which the selected polymer displays a high enough solubility. A representative polymer concentration range in the solvent is between 2 and 10 wt%. For nanoparticle formation by solvent evaporation, the boiling point of the solvent should be below that of the continuous phase, which usually is water or an aqueous solution. Although in the past, the preferred solvents were chlorinated (dichloromethane, chloroform) or aromatic (toluene, benzene), currently there is an increasing trend to replace them by other solvents that show lower toxicity, such as ethyl acetate, ethyl formate, and butyl acetate. Surfactants are usually employed for the stabilization of nano-emulsions. Frequently used emulsifiers in pharmaceutical products are polyoxyethylene castor oil derivatives, polyoxyethylene sorbitan fatty acid esters, polyoxyethylene alkyl ethers, etc. These have been used in pharmaceutical formulations as they are considered to be essentially nontoxic and nonirritant by most administration routes, although generally special caution should be exercised when taken by the parenteral route [39]. For example, polyoxyl 35 castor oil has been reported to cause anaphylactic reactions following parenteral administration [40] and most surfactants cause irritation at higher concentrations (e.g., >20%) [39]. Therefore, once the nanoparticles have been obtained from the nano-emulsions, purification steps may be undertaken to remove the surfactant. However, sometimes the surfactant may show desired biological effects. It has been reported that coating of nanoparticles with polysorbate-80 [41] or polyethyleneglycol (PEG) [42], as well as the presence of Pluronic™ unimers [43] enhance the transport through the blood–brain barrier by different mechanisms. It is also well known that coating of hydrophobic colloidal particles with PEG derivatives, such as poloxamers and poloxamines, avoids the deposition of plasma proteins (also called *opsonization*) and hence their removal from blood circulation by macrophages [44]. The composition of the template nano-emulsion can also affect the physicochemical properties of the nanoparticles obtained, such as size, morphology, internal symmetry, surface charge, polarity, and crystallinity. It has been shown that size affects nanoparticle clearance from circulation, as nanoparticles smaller than 20 nm have been reported to be excreted renally [45]. Nanoparticles with sizes between 30 and 150 nm can accumulate in the bone marrow, heart, kidney, and stomach [46, 47], and nanoparticles with sizes in the range of 150–300 nm can accumulate in the liver and spleen [48].

7.4
Ethylcellulose Nano-Emulsions for Nanoparticle Preparation

Polymeric nano-emulsions have been recently obtained in a water/polyoxyethylene 4 sorbitan monolaurate/polymer solution system by the PIC method at a constant temperature of 25 °C [2]. The polymer solution consisted on a volatile solvent, ethyl acetate, containing a 10 wt% of ethylcellulose. Ethyl acetate is a low toxicity [49] and highly polar solvent. Concerning ethylcellulose, it is a water-insoluble, nonionic cellulose derivative approved by the FDA for pharmaceutical applications. Owing to their characteristics, it is suitable for the encapsulation of water-sensitive drugs. In addition, as the whole preparation procedure is performed at room temperature, temperature-labile components may be solubilized in the system. Nano-emulsions were formed in the mentioned system, over quite a wide range of *O/S* ratios, between 30/70 and 70/30 and water contents above 40 wt%. For the majority of applications, low surfactant concentrations are desirable to improve biocompatibility and to avoid interferences upon application. In the system described in this study, nano-emulsions were obtained at O/S ratios up to 70/30. This ratio represents reasonably low surfactant content for a broad range of applications, including pharmaceutical use. Phase inversion along the emulsification path was verified by conductivity measurements. As mentioned earlier, phase inversion, no matter whether achieved by a temperature [18, 19] or a composition change [3, 20, 23, 31], is essential to attain small and homogeneous droplet sizes. Figure 7.1 shows the conductivity curve obtained when diluting with water a 10 wt% ethylcellulose in ethyl acetate/polyoxyethylene 4 sorbitan monolaurate mixture at a 70/30 ratio. In the absence of water, the conductivity is close to zero. On addition of small amount of water, the conductivity is still low indicating a W/O system. With more addition of water, conductivity increases progressively because of an enhanced mobility of the charged species. When the phase inversion region is reached, conductivity undergoes a sharp increase up to a maximum, indicating the shift to O/W systems. If more water is added, conductivity drops because of the dilution of the conducting species that are provided by the impurities present in the surfactant and the polymer. A similar behavior was observed for other O/S ratios that form nano-emulsions in this system. The phase inversion region appeared, generally, at water contents ranging from 20 to 60 wt%. Bell-shaped conductivity curves have also been described earlier for other systems and are typical for systems, in which no electrolytes are added in the diluting solution along the experimental path [50, 51].

The hydrodynamic diameter (Z_{Ave}) of nano-emulsions with an O/S ratio of 70/30 and 90 wt% of water was 216.3 nm with a polydispersity index of 0.109. This small droplet size ensures good kinetic stability against creaming. This was proven by means of accelerated stability tests carried out by measuring the transmitted light as a function of sample height and time. Transmission intensity data confirmed

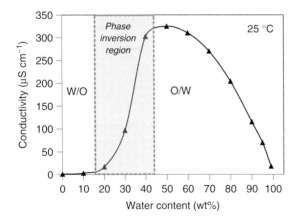

Figure 7.1 Conductivity as a function of water content in the water/polyoxyethylene 4 sorbitan monolaurate/10 wt% ethylcellulose in ethyl acetate system at an O/S ratio of 70/30 at 25 °C.

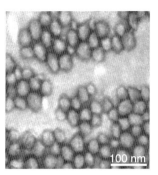

Figure 7.2 Ethylcellulose nanoparticles obtained from the nano-emulsion with an O/S ratio of 70/30, after isolation, washing and concentrating by ultracentrifugation, and finally negative staining with a phosphotungstic acid solution.

the homogeneity and the stability against creaming of the nano-emulsion. This behavior is expected from nano-emulsions because of their small droplet size that makes them less prone to migration phenomena, as described by Stokes' law. These nano-emulsions were used for nanoparticle preparation by the solvent evaporation method. The nanoparticle size, as determined by dynamic light scattering, was around 80 nm. This size is suitable for biomedical applications, and is in good agreement with the diameter expected from theoretical calculations considering that the number of dispersed entities is kept constant, during the evaporation process. Nanoparticles were also characterized by transmission electron microscopy (TEM). For this purpose, the nanoparticles were negatively stained with a 2 wt% phosphotungstic acid solution. The nanoparticles show a globular shape (Figure 7.2).

The mean particle size obtained from image analysis of TEM micrographs was about 44 nm. This size is smaller than that provided by the light scattering technique, because the latter provides the hydrodynamic diameter while measurements performed by TEM image analysis are based on the hard sphere diameter. Furthermore, a slight shrinkage due to drying of the TEM sample needed for observation may play a role. The nanoparticle dispersion exhibits a Schulz-like monomodal distribution, as displayed in Figure 7.3.

7.5
Final Remarks

Polymeric nano-emulsions obtained by the PIC method can be used as templates for nanoparticle preparation. The selection of the components as well as the preparation method determines the properties of the nano-emulsion (droplet size, stability, etc.) as well as its suitability for biomedical applications (surfactant type and concentration, low toxicity of the solvent, polymer nature, and molecular weight, etc.). Ethylcellulose nanoparticles have been obtained from nano-emulsions by the PIC method, at a relatively high oil-to-surfactant ratio, using a low-toxicity

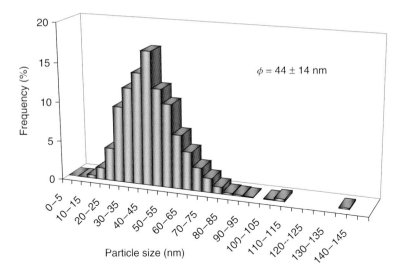

Figure 7.3 Nanoparticle size distribution as determined by transmission electron microscopy image analysis.

solvent, an FDA approved polymer, and room temperature. Therefore, they are suitable candidates for biomedical applications.

Acknowledgments

Financial support from Spanish Ministry of Economy and Competitiveness, MEC (grant CTQ2011-29336-C03-01), and Generalitat de Catalunya (grant 2009SGR-961) is acknowledged. CIBER-BBN is an initiative funded by the VI National R&D&i Plan 2008–2011, Iniciativa Ingenio 2010, Consolider Program, CIBER Actions, and financed by the Instituto de Salud Carlos III with assistance from the European Regional Development Fund.

References

1. Solans, C., Izquierdo, P., Nolla, J., Azemar, N., and Garcia-Celma, M.J. (2005) *Curr. Opin. Colloid Interface Sci.*, **10** (3–4), 102–110.
2. Calderó, G., García-Celma, M.J., and Solans, C. (2011) *J. Colloid Interface Sci.*, **353**, 406–411.
3. Sadurní, N., Solans, C., Azemar, N., and García-Celma, M.J. (2005) *Eur. J. Pharm. Sci.*, **26**, 438–445.
4. Kreuter, J. (2007) *Int. J. Pharm.*, **331**, 1–10.
5. Wagner, V., Hüsing, B., Gaisser, S., and Bock, A.-K. (2008) Nanomedicine: Drivers for Development and Possible Impacts. Report No. 46744, European Commission's (EC) Joint Research Center (JRC), pp. 1–116, http://ipts.jrc.ec.europa.eu.
6. Camino, N.A. and Pilosof, A.M.R. (2011) *Food Hydrocoll.*, **25**, 1051–1062.
7. Delmas, Th., Piraux, H., Couffin, A.-C., Texier, I., Vinet, F., Poulin, Ph., Cates,

M.E., and Bibette, J. (2011) *Langmuir*, **27** (5), 1683–1692.

8. Kentish, S., Wooster, T.J., Ashokkumar, M., Balachandran, S., Mawson, R., and Simons, L. (2008) *Innov. Food Sci. Emerg. Technol.*, **9**, 170–175.
9. Qian, Ch. and McClements, D.J. (2011) *Food Hydrocoll.*, **25**, 1000–1008.
10. Ganachaud, F. and Katz, J.L. (2005) *ChemPhysChem*, **6**, 209–216.
11. Pouton, C.W. (1997) *Adv. Drug Delivery Rev.*, **25**, 47–58.
12. Gursoy, R.N. and Benita, S. (2004) *Biomed. Pharmacother.*, **58**, 173–182.
13. Morales, D., Gutiérrez, J.M., García-Celma, M.J., and Solans, C. (2003) *Langmuir*, **19**, 7196–7200.
14. Cheng, S., Guo, Y., and Zetterland, P.B. (2010) *Macromolecules*, **43**, 7905–7907.
15. Solè, I., Maestro, A., Pey, C.M., González, C., Solans, C., and Gutiérrez, J.M. (2006) *Eng. Aspects*, **288**, 138–143.
16. Forgiarini, A., Esquena, J., Gonzalez, C., and Solans, C. (2002) *J. Disper. Sci. Technol.*, **23**, 209–217.
17. Shinoda, K. and Saito, H. (1968) *J. Colloid Interface Sci.*, **26**, 70–74.
18. Saito, H. and Shinoda, K. (1969) *J. Colloid Interface Sci.*, **30** (2), 258–263.
19. Wadle, A., Förster, Th., and von Rybinski, W. (1993) *Colloids Surf., A: Physicochem. Eng. Aspects*, **76**, 51–57.
20. Forgiarini, A., Esquena, J., González, C., and Solans, C. (2001) *Langmuir*, **17**, 2076–2083.
21. Pey, C.M., Maestro, A., Solé, I., González, C., Solans, C., and Gutiérrez, J.M. (2006) *Colloids Surf., A: Physicochem. Eng. Aspects*, **288**, 144–150.
22. Wang, L., Mutch, K.J., Eastoe, J., Heenan, R.K., and Dong, J. (2008) *Langmuir*, **24**, 6092–6099.
23. Solè, I., Pey, C.M., Maestro, A., González, C., Porras, M., and Solans, C. (2010) *J. Colloid Interface Sci.*, **344**, 417–423.
24. Porras, M., Solans, C., González, C., and Gutiérrez, J.M. (2008) *Colloids Surf., A: Physicochem. Eng. Aspects*, **324**, 181–188.
25. Usón, N., García, M.J., and Solans, C. (2004) *Colloids Surf., A: Physicochem. Eng. Aspects*, **250** (1–3), 415–421.
26. Peng, L.C., Liu, C.H., Kwan, C.C., and Huang, K.-F. (2010) *Colloids Surf., A: Physicochem. Eng. Aspects*, **370**, 136–142.
27. Solè, I., Maestro, A., González, C., Solans, C., and Gutiérrez, J.M. (2006) *Langmuir*, **22**, 8326–8332.
28. Lamaallam, S., Bataller, H., Dicharry, C., and Lachaise, J. (2005) *Colloids Surf., A: Physicochem. Eng. Aspects*, **270–271**, 44–51.
29. Dicharry, C., Bataller, H., Lamallam, S., Lachaise, J., and Graciaa, A. (2003) *J. Disper. Sci. Technol.*, **24**, 237–248.
30. Hessien, M., Singh, N., Kim, Ch.H., and Prouzet, E. (2011) *Langmuir*, **27**, 2299–2307.
31. Fernandez, P., André, V., Rieger, J., and Kühnle, A. (2004) *Colloids Surf., A: Physicochem. Eng. Aspects*, **251**, 53–58.
32. Jahanzad, F., Josephides, D., Monsourian, A., and Saijjadi, Sh. (2010) *Ind. Eng. Chem. Res.*, **49**, 7631–7637.
33. Sagitani, H. (1981) *J. Am. Oil Chem. Soc.*, **58**, 738–743.
34. Seymour, L.W., Duncan, R., Strohalm, J., and Kopecek, J. (1987) *J. Biomed. Mater. Res.*, **21**, 1341–1358.
35. Seymour, L.W., Miyamoto, Y., Maeda, H., Brereton, M., Strohalm, J., Ulbrich, K., and Duncan, R. (1995) *Eur. J. Cancer*, **31**, 766–770.
36. Robinson, B.V., Sullivan, F.M., Borzelleca, J.F., and Schwartz, S.L. (1990) *PVP: A Critical Review of the Kinetics and Toxicology of Polyvinylpyrrolidone (Povidone)*, Lewis Publishers, Inc., Chelsea, MI.
37. Brunner, S., Fürtbauer, E., Sauer, T., Kursa, M., and Wagner, E. (2002) *Mol. Ther.*, **5**, 80–86.
38. Talcott, B. and Moore, M.S. (1999) *Trends Cell Biol.*, **9**, 312–318.
39. Rowe, R.C., Sheskey, P.J., and Owen, S.C. (eds) (1986) in *Handbook of Pharmaceutical Excipients*, 5th edn, Pharmaceutical Press, London, pp. 572–579.
40. Dye, D. and Watkins, J. (1980) *Br. Med. J.*, **280**, 1353.
41. Kreuter, J., Ramge, P., Petrov, V., Hamm, S., Gelperina, S.E., Engelhardt, B., Alyautdin, R., von Briesen, H., and Begley, D.J. (2003) *Pharm. Res.*, **20**, 409–416.

42. Calvo, P., Gouritin, B., Villarroya, H., Eclancher, F., Giannavola, C., Klein, C., Andreux, J.P., and Couvreur, P. (2002) *Eur. J. Neurosci.*, **15**, 1317–1326.
43. Batrakova, E.V., Li, S., Vinogradov, S.V., Alakhov, V.Y., Miller, D.W., and Kabanov, A.V. (2001) *J. Pharmacol. Exp. Ther.*, **299**, 483–493.
44. Peracchia, M.T., Harmisch, S., Pinto-Alphandary, H., Gulik, A., Dedieu, J.C., Desmaele, D., D'Angelo, J., Muller, R.H., and Couvreur, P. (1999) *Biomaterials*, **20**, 1269–1275.
45. Choi, H.S., Liu, W., Misra, P., Tanaka, E., Zimmer, J.P., Ipe, B.I., Bawendi, M.G., and Frangioni, J.V. (2007) *Nat. Biotechnol.*, **25**, 1165–1170.
46. Banerjee, T., Mitra, S., Singh, A.K., Sharma, R.K., and Maitra, A. (2002) *Int. J. Pharm.*, **243**, 93–105.
47. Moghimi, S.M. (1995) *Adv. Drug Delivery Rev.*, **17**, 61–73.
48. Moghimi, S.M. (1995) *Adv. Drug Delivery Rev.*, **17**, 103–115.
49. (2005) International Conference on Harmonisation of Technical Requirements for Registration of Pharmaceuticals for Human Use. Impurities: Guideline for Residual Solvents ICH Harmonised Tripartite Guideline.
50. Meziani, A., Zradba, A., Touraud, D., Clausse, M., and Kunz, W. (1997) *J. Mol. Liq.*, **73–74**, 107–118.
51. Meziani, A., Zradba, A., Touraud, D., Clausse, M., and Kunz, W. (2000) *J. Mol. Liq.*, **84**, 301–311.

8
Rheology and Stability of Sterically Stabilized Emulsions
Tharwat F. Tadros

8.1
Introduction

Polymeric surfactants are essential materials for the preparation and stabilization of many emulsions, of which we mention agrochemicals, pharmaceuticals, and personal care products [1]. One of the most important applications of polymeric surfactants is in the preparation of oil-in-water (O/W) and water-in-oil (W/O) emulsions [2, 3]. In this case, the hydrophobic portion of the surfactant molecule should adsorb "strongly" at the O/W or become dissolved in the oil phase, leaving the hydrophilic components in the aqueous medium, whereby they become strongly solvated by the water molecules.

For stabilization of emulsions against flocculation, coalescence, and Ostwald ripening the following criteria must be satisfied [4, 5]: (i) complete coverage of the droplets by the surfactant – any bare patches may result in flocculation as a result of van der Waals attraction or bridging; (ii) strong adsorption (or "anchoring") of the surfactant molecule to the surface of droplet; (iii) strong solvation (hydration) of the stabilizing chain to provide effective steric stabilization; (iv) reasonably thick adsorbed layer to prevent weak flocculation [1].

Most of the above criteria for stability are best served by using a polymeric surfactant [2]. In particular, molecules of the A-B, A-B-A blocks and BA_n (or AB_n) grafts (see following text) are the most efficient for stabilization of emulsions. In this case, the B chain (referred to as the *anchoring* chain) is chosen to be highly insoluble in the medium and with a high affinity to the surface or soluble in the oil. The A chain is chosen to be highly soluble in the medium and strongly solvated by its molecules. These block and graft copolymers are ideal for preparation of concentrated emulsions, which are needed in many industrial applications.

This chapter starts with a section on the general classification of polymeric surfactants. The next section is devoted to the principles involved in stabilization of emulsions using polymeric surfactants, that is, the general theory of steric stabilization. Two sections are devoted to describe the use of polymeric surfactants for stabilization of emulsions. The principles of the various rheological techniques that are applied in this study are briefly described. The applications of the various rheological techniques for studying the interactions in sterically stabilized emulsions are described in subsequent sections. Two types of emulsions are considered, namely, O/W and W/O that are stabilized using block copolymer emulsifiers.

8.2
General Classification of Polymeric Surfactants

Perhaps the simplest type of a polymeric surfactant is a homopolymer, which is formed from the same repeating units such as poly(ethylene oxide) (PEO) or poly(vinyl pyrrolidone). These homopolymers have little surface activity at the O/W interface, as the homopolymer segments (ethylene oxide or vinylpyrrolidone) are highly water soluble and have little affinity to the interface. Even if the adsorption energy per monomer segment to the surface is small (fraction of kT, where k is the Boltzmann constant and T is the absolute temperature), the total adsorption energy per molecule may be sufficient to overcome the unfavorable entropy loss of the molecule at the S/L interface.

Clearly, homopolymers are not the most suitable emulsifiers. A small variant is to use polymers that contain specific groups that have high affinity to the oil. This is exemplified by partially hydrolyzed poly(vinyl acetate) (PVAc), technically referred to as poly(vinyl alcohol) (PVA). The polymer is prepared by partial hydrolysis of PVAc, leaving some residual vinyl acetate groups. Most commercially available PVA molecules contain 4–12% acetate groups. These acetate groups that are hydrophobic give the molecule its amphipathic character. On a hydrophobic surface such as oil, the polymer adsorbs with preferential attachment to the acetate groups on the surface, leaving the more hydrophilic vinyl alcohol segments dangling in the aqueous medium. These partially hydrolyzed PVA molecules also exhibit surface activity at the O/W interface [6].

The most convenient polymeric surfactants are those of the block and graft copolymer type. A block copolymer is a linear arrangement of blocks of variable monomer composition. The nomenclature for a diblock is poly-A-*block*-poly-B and for a triblock it is poly-A-*block*-poly-B-poly-A. One of the most widely used triblock polymeric surfactants are the "Pluronics" (BASF, Germany), which consist of two poly-A blocks of PEO and one block of poly(propylene oxide) (PPO). Several chain lengths of PEO and PPO are available. More recently, triblocks of PPO-PEO-PPO (inverse Pluronics) became available for some specific applications [6].

The above polymeric triblocks can be applied as emulsifiers, whereby the assumption is made that the hydrophobic PPO chain resides at the hydrophobic surface, leaving the two PEO chains dangling in aqueous solution, hence providing steric repulsion.

Although the above triblock polymeric surfactants have been widely used in various applications in emulsions, some doubt has arisen on how effective these can be. It is generally accepted that the PPO chain is not sufficiently hydrophobic to provide a strong "anchor" to an oil droplet. Indeed, the reason for the surface activity of the PEO-PPO-PEO triblock copolymers at the O/W interface may stem from a process of "rejection" anchoring of the PPO chain, as it is not soluble in both oil and water.

Several other di- and triblock copolymers have been synthesized, although these are of limited commercial availability. Typical examples are diblocks of polystyrene-*block*-polyvinyl alcohol, triblocks of poly(methyl methacrylate)-*block*-poly(ethylene oxide)-*block*-poly(methyl methacrylate), diblocks of polystyrene-*block*-polyethylene oxide, and triblocks of polyethylene oxide-*block*-polystyrene-polyethylene oxide [6].

An alternative (and perhaps more efficient) polymeric surfactant is the amphipathic graft copolymer consisting of a polymeric backbone B (polystyrene or poly(methyl methacrylate)) and several A chains ("teeth") such as PEO [6]. This graft copolymer is sometimes referred to as a *"comb" stabilizer*. This copolymer is usually prepared by grafting a macromonomer such as methoxy polyethylene oxide methacrylate with polymethyl methacrylate. The "grafting onto" technique has also been used to synthesize polystyrene-polyethylene oxide graft copolymers.

Recently, graft copolymers based on polysaccharides [7–9] have been developed for stabilization of emulsions. One of the most useful graft copolymers are those based on inulin that is obtained from chicory roots. It is a linear polyfructose chain with a glucose end. When extracted from chicory roots, inulin has a wide range of chain lengths ranging from 2 to 65 fructose units. It is fractionated to obtain a molecule with narrow molecular weight distribution with a degree of polymerization >23 and this is commercially available as INUTEC®N25. The latter molecule is used to prepare a series of graft copolymers by random grafting of alkyl chains (using alkyl isocyanate) on the inulin backbone. The first molecule of this series is INUTEC® SP1 (Beneo–Remy, Belgium), which is obtained by random grafting of C_{12} alkyl chains. It has an average molecular weight of ∼5000 Da and its structure is given in Figure 8.1. The molecule is schematically illustrated in

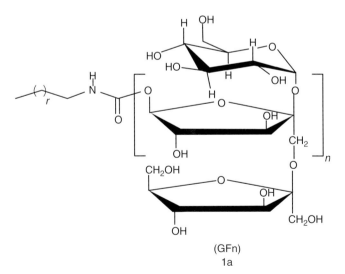

Figure 8.1 Structure of INUTEC® SP1.

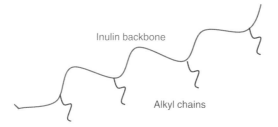

Figure 8.2 Schematic representation of INUTEC®SP1 polymeric surfactant.

Figure 8.2, which shows the hydrophilic polyfructose chain (backbone) and the randomly attached alkyl chains.

The main advantages of INUTEC SP1 as stabilizer for emulsions are (i) strong adsorption to the droplet by multipoint attachment with several alkyl chains, which ensures lack of desorption and displacement of the molecule from the interface; (ii) strong hydration of the linear polyfructose chains both in water and in the presence of high electrolyte concentrations and high temperatures, which ensures effective steric stabilization (see below).

8.3
Interaction between Droplets Containing Adsorbed Polymeric Surfactant Layers: Steric Stabilization

When two droplets each with a radius R and containing an adsorbed polymer layer with a hydrodynamic thickness δ_h, approach each other to a surface–surface separation distance h that is smaller than $2\delta_h$, the polymer layers interact with each other resulting in two main situations [10]: (i) the polymer chains may overlap with each other and (ii) the polymer layer may undergo some compression. In both cases, there is an increase in the local segment density of the polymer chains in the interaction region. This is schematically illustrated in Figure 8.3. The real situation is perhaps in between the above two cases, that is, the polymer chains may undergo some interpenetration and some compression.

Provided the dangling chains (the A chains in A-B, A-B-A block, or BA$_n$ graft copolymers) are in a good solvent, this local increase in segment density in the

Interpenetration without compression

Compression without interpenetration

Figure 8.3 Schematic representation of the interaction between droplets containing adsorbed polymer layers.

interaction zone will result in strong repulsion as a result of two main effects: (i) Increase in the osmotic pressure in the overlap region as a result of the unfavorable mixing of the polymer chains, when these are in good solvent conditions [10]. This is referred to as *osmotic repulsion* or *mixing interaction* and it is described by a free energy of interaction G_{mix}. (ii) Reduction of the configurational entropy of the chains in the interaction zone [10]; this entropy reduction results from the decrease in the volume available for the chains when these are either overlapped or compressed. This is referred to as *volume restriction interaction, entropic* or *elastic interaction* and it is described by a free energy of interaction G_{el}.

Combination of G_{mix} and G_{el} is usually referred to as the *steric interaction free energy*, G_s, that is,

$$G_s = G_{mix} + G_{el} \tag{8.1}$$

The sign of G_{mix} depends on the solvency of the medium for the chains. If in a good solvent, that is, the Flory–Huggins interaction parameter χ is less than 0.5, then G_{mix} is positive and the mixing interaction leads to repulsion (see following text). In contrast, if $\chi > 0.5$ (i.e., the chains are in a poor solvent condition), G_{mix} is negative and the mixing interaction becomes attractive. G_{el} is always positive and hence in some cases one can produce stable dispersions in a relatively poor solvent (enhanced steric stabilization).

8.3.1
Mixing Interaction G_{mix}

This results from the unfavorable mixing of the polymer chains, when these are in a good solvent condition. This is schematically shown in Figure 8.4.

Consider two spherical droplets with the same radius, each containing an adsorbed polymer layer with thickness δ. Before overlap, one can define in each polymer layer a chemical potential for the solvent μ_i^α and a volume fraction for the polymer in the layer ϕ_2. In the overlap region (volume element dV), the chemical

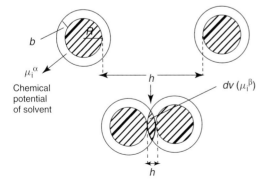

Figure 8.4 Schematic representation of polymer layer overlap.

potential of the solvent is reduced to μ_i^β. This results from the increase in polymer segment concentration in this overlap region.

In the overlap region, the chemical potential of the polymer chains is now higher than in the rest of the layer (with no overlap). This amounts to an increase in the osmotic pressure in the overlap region; as a result, solvent will diffuse from the bulk to the overlap region, thus, separating the particles, and hence, a strong repulsive energy arises from this effect. The above repulsive energy can be calculated by considering the free energy of mixing of two polymer solutions, as, for example, treated by Flory and Krigbaum [11]. The free energy of mixing is given by two terms (i) an entropy term that depends on the volume fraction of and solvent and (ii) an energy term that is determined by the Flory–Huggin interaction parameter χ.

Using the above theory, one can derive an expression for the free energy of mixing of two polymer layers (assuming a uniform segment density distribution in each layer) surrounding two spherical particles as a function of the separation distance h between the particles. The expression for G_{mix} is

$$\frac{G_{\mathrm{mix}}}{kT} = \left(\frac{2V_2^2}{V_1}\right) v_2^2 \left(\frac{1}{2} - \chi\right)\left(\delta - \frac{h}{2}\right)^2 \left(3R + 2\delta + \frac{h}{2}\right) \quad (8.2)$$

where k is the Boltzmann constant, T is the absolute temperature, V_2 is the molar volume of polymer, V_1 is the molar volume of solvent, and v_2 is the number of polymer chains per unit area.

The sign of G_{mix} depends on the value of the Flory–Huggins interaction parameter χ: if $\chi < 0.5$, G_{mix} is positive and the interaction is repulsive; if $\chi > 0.5$, G_{mix} is negative and the interaction is attractive; if $\chi = 0.5$, $G_{\mathrm{mix}} = 0$ and this defines the θ-condition.

8.3.2
Elastic Interaction G_{el}

This arises from the loss in configurational entropy of the chains on the approach of a second droplet. As a result of this approach, the volume available for the chains becomes restricted, resulting in loss of the number of configurations. This can be illustrated by considering a simple molecule, represented by a rod that rotates freely in a hemisphere across a surface (Figure 8.5). When the two surfaces are separated

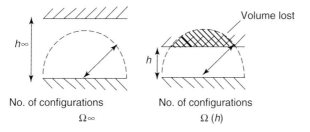

Figure 8.5 Schematic representation of configurational entropy loss on approach of a second particle.

8.3 Interaction between Droplets Containing Adsorbed Polymeric Surfactant Layers

by an infinite distance, ∞, the number of configurations of the rod is $\Omega(\infty)$, which is proportional to the volume of the hemisphere. When a second particle approaches a distance h such that it cuts the hemisphere (losing some volume), the volume available to the chains is reduced and the number of configurations becomes $\Omega(h)$, which is less than $\Omega(\infty)$. For two flat plates, G_{el}, is given by the following expression:

$$\frac{G_{el}}{kT} = 2\nu_2 \ln\left[\frac{\Omega(h)}{\Omega(\infty)}\right] = 2\nu_2 R_{el}(h) \qquad (8.3)$$

where $R_{el}(h)$ is a geometric function whose form depends on the segment density distribution. It should be stressed that G_{el} is always positive and could play a major role in steric stabilization. It becomes very strong when the separation distance between the particles becomes comparable to the adsorbed layer thickness δ.

Combination of G_{mix} and G_{el} with G_A (van der Waals attraction) gives the total energy of interaction G_T (assuming there is no contribution from any residual electrostatic interaction), that is,

$$G_T = G_{mix} + G_{el} + G_A \qquad (8.4)$$

A schematic representation of the variation of G_{mix}, G_{el}, G_A, and G_T with surface–surface separation distance h is shown in Figure 8.6.

G_{mix} increases very sharply with decrease of h, when $h < 2\delta$. G_{el} increases very sharply with decrease of h, when $h < \delta$. G_T versus h shows a minimum, G_{min}, at separation distances comparable to 2δ. When $h < 2\delta$, G_T shows a rapid increase with decrease in h.

The depth of the minimum depends on the Hamaker constant A, the droplet radius R, and adsorbed layer thickness δ. G_{min} increases with increase of A and R. At a given A and R, G_{min} increases with decrease in δ (i.e., with decrease of the molecular weight, M_w, of the stabilizer). This is illustrated in Figure 8.7, which shows the energy–distance curves as a function of δ/R. The larger the value of δ/R, the smaller the value of G_{min}. In this case, the system may approach thermodynamic stability, as is the case with nanoemulsions.

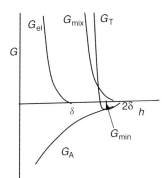

Figure 8.6 Energy–distance curves for sterically stabilized systems.

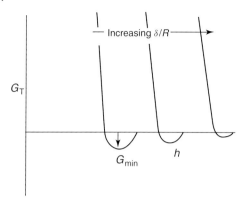

Figure 8.7 Variation of G_{min} with δ/R.

8.4
Emulsions Stabilized by Polymeric Surfactants

As mentioned above, the most effective method for emulsion stabilization is to use polymeric surfactants that strongly adsorb at the O/W or W/O interface and produce effective steric stabilization against strong flocculation, coalescence and Ostwald ripening [12].

As mentioned above, a graft copolymer of the AB_n type was synthesized by grafting several alkyl groups on an inulin (polyfructose) chain. The polymeric surfactant (INUTEC SP1) consists of a linear polyfructose chain (the stabilizing A chain) and several alkyl groups (the B chains) that provide multianchor attachment to the oil droplets. This polymeric surfactant produces enhanced steric stabilization both in water and high electrolyte concentrations.

For W/O emulsions an A-B-A block copolymer of poly(12-hydroxystearic acid) (PHS) (the A chains) and PEO (the B chain): PHS-PEO-PHS was used. The PEO chain (which is soluble in the water droplets) forms the anchor chain, whereas the PHS chains form the stabilizing chains. PHS is highly soluble in most hydrocarbon solvents and is strongly solvated by its molecules. The structure of the PHS-PEO-PHS block copolymer is schematically shown in Figure 8.8.

The conformation of the polymeric surfactant at the W/O interface is schematically shown in Figure 8.9.

Emulsions of Isopar M/water and cyclomethicone/water were prepared using INUTEC SP1. 50/50 (v/v) O/W emulsions were prepared and the emulsifier concentration was varied from 0.25 to 2 (w/v)% based on the oil phase. 0.5 (w/v)% emulsifier was sufficient for stabilization of these 50/50 (v/v) emulsions [12].

The emulsions were stored at room temperature and 50°C and optical micrographs were taken at intervals of time (for a year), in order to check the stability. Emulsions prepared in water were very stable showing no change in droplet size distribution over more than a year and this indicated absence of coalescence. Any weak flocculation that occurred was reversible and the

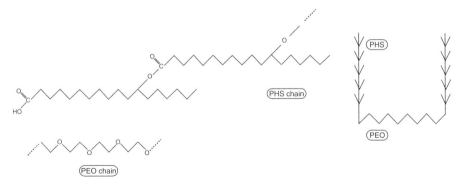

Figure 8.8 Schematic representation of the structure of PHS-PEO-PHS block copolymer.

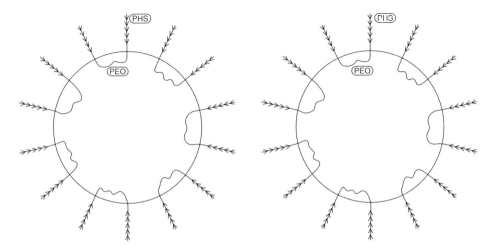

Figure 8.9 Conformation of PHS-PEO-PHS polymeric surfactant at the W/O interface.

emulsion could be redispersed by gentle shaking. Figure 8.10 shows an optical micrograph for a dilute 50/50 (v/v) emulsion that was stored for 1.5 and 14 weeks at 50 °C.

No change in droplet size was observed after storage for more than one year at 50 °C, indicating absence of coalescence. The same result was obtained when using different oils. Emulsions were also stable against coalescence in the presence of high electrolyte concentrations (up to 4 mol dm^{-3} or ~25% NaCl).

The above stability in high electrolyte concentrations is not observed with polymeric surfactants based on PEO.

The high stability observed using INUTEC SP1 is related to its strong hydration, both in water and in electrolyte solutions. The hydration of inulin (the backbone of HMI (hydrophobically modified inulin)) could be assessed using cloud point measurements. A comparison was also made with PEO with two molecular weights, namely, 4000 and 20 000 Da.

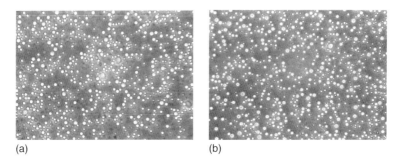

(a) (b)

Figure 8.10 Optical micrographs of O/W emulsions stabilized with INUTEC® SP1 stored at 50 °C for 1.5 weeks (a) and 14 weeks (b).

Solutions of PEO 4000 and 20 000 showed a systematic decrease of cloud point with increase in NaCl or $MgSO_4$ concentration. In contrast, inulin showed no cloud point up to 4 mol dm^{-3} NaCl and up to 1 mol dm^{-3} $MgSO_4$.

The above results explain the difference between PEO and inulin. With PEO, the chains show dehydration when the NaCl concentration is increased above 2 mol dm^{-3} or $MgSO_4$ is increased above 0.5 mol dm^{-3}. The inulin chains remain hydrated at much higher electrolyte concentrations. It seems that the linear polyfructose chains remain strongly hydrated at high temperature and high electrolyte concentrations.

The high-emulsion stability obtained when using INUTEC SP1 can be accounted for by the following factors: (i) the multipoint attachment of the polymer by several alkyl chains that are grafted on the backbone; (ii) the strong hydration of the polyfructose "loops" in both water and high electrolyte concentrations (χ remains below 0.5 under these conditions); (iii) the high volume fraction (concentration) of the loops at the interface; (iv) enhanced steric stabilization, which is the case with multipoint attachment that produces strong elastic interaction.

Evidence for the high stability of the liquid film between emulsion droplets when using INUTEC SP1 was obtained by Exerowa $et\ al.$ [13] using disjoining pressure measurements. This is illustrated in Figure 8.11, which shows a plot of disjoining pressure versus separation distance between two emulsion droplets at various electrolyte concentrations. The results show that by increasing the capillary pressure, a stable Newton Black Film (NBF) is obtained at a film thickness of $\sim$7 nm. The lack of rupture of the film at the highest pressure applied of 4.5×10^4 Pa indicates the high stability of the film in water and in high electrolyte concentrations (up to 2.0 mol dm^{-3} NaCl).

The lack of rupture of the NBF up to the highest pressure applied, namely, 4.5×10^4 Pa clearly indicates the high stability of the liquid film in the presence of high NaCl concentrations (up to 2 mol dm^{-3}). This result is consistent with the high-emulsion stability obtained at high electrolyte concentrations and high temperature. Emulsions of Isopar M-in-water are very stable under such conditions and this could be accounted for by the high stability of the NBF. The droplet size of 50 : 50 O/W emulsions prepared using 2% INUTEC SP1 is in the region of

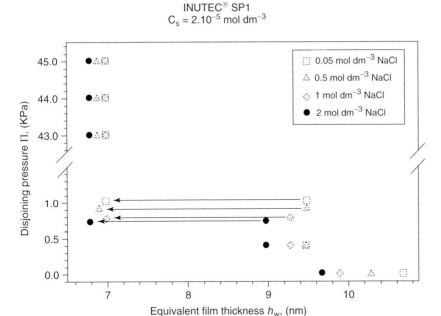

Figure 8.11 Variation of disjoining pressure with equivalent film thickness at various NaCl concentrations.

1–10 µm. This corresponds to a capillary pressure of $\sim 3 \times 10^4$ Pa for the 1 µm drops and $\sim 3 \times 10^3$ Pa for the 10 µm drops. These capillary pressures are lower than those to which the NBF have been subjected to and this clearly indicates the high stability obtained against coalescence in these emulsions.

8.4.1
W/O Emulsions Stabilized with PHS-PEO-PHS Block Copolymer

W/O emulsions (the oil being Isopar M) were prepared using PHS-PEO-PHS block copolymer at high water volume fractions (>0.7). The emulsions have a narrow droplet size distribution with a z-average radius of 183 nm. They also remained fluid up to high water volume fractions (>0.6). This could be illustrated from viscosity–volume fraction curves as is shown in Figure 8.12.

The effective volume fraction ϕ_{eff} of the emulsions (the core droplets plus the adsorbed layer) could be calculated from the relative viscosity and by using the Dougherty–Krieger equation [14],

$$\eta_r = \left[1 - \frac{\phi_{\text{eff}}}{\phi_p}\right]^{-[\eta]\phi_p} \tag{8.5}$$

where η_r is the relative viscosity, ϕ_p is the maximum packing fraction (~ 0.7), and $[\eta]$ is the intrinsic viscosity that is equal to 2.5 for hard spheres.

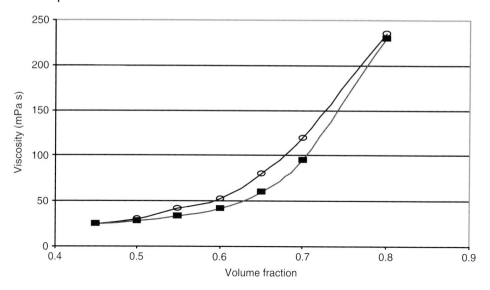

Figure 8.12 Viscosity–volume fraction for W/O emulsion stabilized with PHS-PEO-PHS block copolymer.

The calculations based on Eq. (8.5) are shown in Figure 8.12 (square symbols). From the effective volume fraction ϕ_{eff} and the core volume fraction ϕ, the adsorbed layer thickness could be calculated. This was found to be in the region of 10 nm at $\phi = 0.4$ and it decreased with increase in ϕ.

The W/O emulsions prepared using the PHS-PEO-PHS block copolymer remained stable both at room temperature and at 50 °C. This is consistent with the structure of the block copolymer: the B chain (PEO) is soluble in water and it forms a very strong anchor at the W/O interface. The PHS chains (the A chains) provide effective steric stabilization as the chains are highly soluble in Isopar M and are strongly solvated by its molecules.

8.5
Principles of Rheological Techniques

8.5.1
Steady State Measurements

In this case, the emulsion that is placed in the gap between concentric cylinders, cone and plate or parallel plate are subjected to constant shear rate until a steady state is reached, whereby the stress remains constant. The measurement allows one to obtain the stress and viscosity as a function of shear rates [15]. Most emulsions that have high volume fraction ϕ of the disperse phase ($\phi > 0.1$), and contain "thickeners" to reduce creaming or sedimentation, show non-Newtonian flow as illustrated in Figure 8.13.

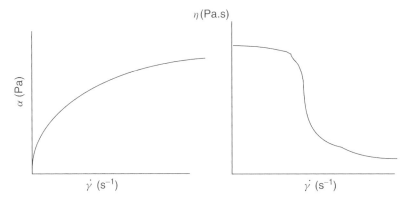

Figure 8.13 Pseudoplastic (shear thinning) system.

Various models can be used to analyze the above flow curves, of which the Bingham, Power law, and Herscel–Bulkley are the most commonly used.

8.5.1.1 Bingham Plastic Systems

$$\sigma = \sigma_\beta + \eta_{pl}\,\dot\gamma \tag{8.6}$$

The system shows a (dynamic) yield stress σ_β that can be obtained by extrapolation to zero shear rate [16]. Clearly, at and below σ_β the viscosity $\eta \to \infty$. The slope of the linear curve gives the plastic viscosity η_{pl}.

8.5.1.2 Pseudoplastic (Shear Thinning) System

In this case, the system does not show a yield value. It shows a limiting viscosity $\eta(0)$ at low shear rates (that is referred to as *residual* or *zero shear viscosity*). The flow curve can be fitted to a power law fluid model (Ostwald de Waele)

$$\sigma = k\dot\gamma^n \tag{8.7}$$

where k is the consistency index and n is the shear thinning index; $n < 1$.

By fitting the experimental data to Eq. (8.7), k and n can be obtained. The viscosity at a given shear rate can be calculated as

$$\eta = \frac{\sigma}{\dot\gamma} = \frac{k\dot\gamma^n}{\dot\gamma} = k\dot\gamma^{n-1} \tag{8.8}$$

The power law model (Eq. (8.7)) fits the experimental results for many non-Newtonian systems over two or three decades of shear rate. Thus, this model is more versatile than the Bingham model, although care should be taken in applying this model outside the range of data used to define it. In addition, the power law fluid model fails at high shear rates, whereby the viscosity must ultimately reach a constant value, that is, the value of n should approach unity.

8.5.1.3 Herschel–Bulkley General Model

Many systems show a dynamic yield value followed by a shear thinning behavior [17]. The flow curve can be analyzed using the Herschel–Bulkley equation:

$$\sigma = \sigma_\beta + k\dot{\gamma}^n \tag{8.9}$$

When $\sigma_\beta = 0$, Eq. (8.14) reduces to the Power fluid model. When $n = 1$, Eq. (8.9) reduces to the Bingham model. When $\sigma_\beta = 0$ and $n = 1$, Eq. (8.15) becomes the Newtonian equation. The Herschel–Bulkley equation fits most flow curves with a good correlation coefficient and hence it is the most widely used model.

8.5.2
Constant Stress (Creep) Measurements

A constant stress σ is applied on the system (that may be placed in the gap between two concentric cylinders or a cone and plate geometry) and the strain (relative deformation) γ or compliance $J (= \gamma/\sigma,\ \text{Pa}^{-1})$ is followed as a function of time for a period of t. At $t = t$, the stress is removed and the strain γ or compliance J is followed for another period t [15].

The above procedure is referred to as *"creep measurement"*. From the variation of J with t when the stress is applied and the change of J with t when the stress is removed (in this case J changes sign), viscoelastic response can be described as is illustrated in Figure 8.14.

For a viscoelastic liquid, whereby the compliance $J(t)$ is given by two components an elastic component J_e that is given by the reciprocal of the instantaneous modulus

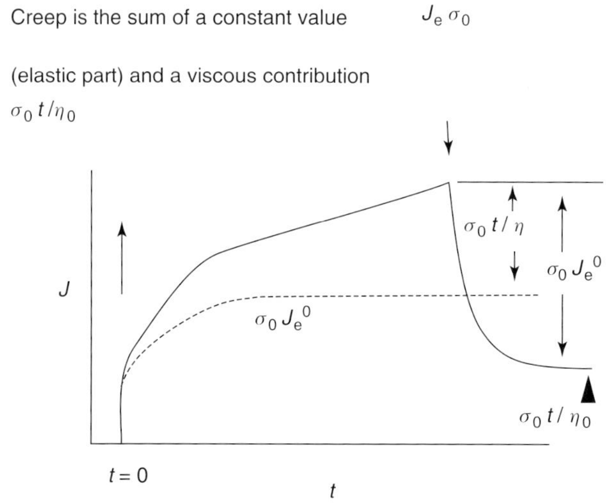

Creep is the sum of a constant value $\quad J_e\,\sigma_0$

(elastic part) and a viscous contribution $\sigma_0 t/\eta_0$

Figure 8.14 Creep curve for a viscoelastic liquid.

and a viscous component J_v that is given by $t/\eta(0)$,

$$J(t) = \frac{1}{G(0)} + \frac{t}{\eta(0)} \tag{8.10}$$

Figure 8.14 also shows the recovery curve which gives $\sigma_0 J_e^0$ and when this is subtracted from the total compliance gives $\sigma_0 t/\eta(0)$.

The driving force for relaxation is spring and the viscosity controls the rate. The Maxwell relaxation time τ_M is given by

$$\tau_M = \frac{\eta(0)}{G(0)} \tag{8.11}$$

8.5.3
Dynamic (Oscillatory) Measurements

This is the response of the material to an oscillating stress or strain [15]. When a sample is constrained in, say, a cone and plate or concentric cylinder assembly, an oscillating strain at a given frequency $\omega(\text{rad s}^{-1})$ ($\omega = 2v\pi$, where v is the frequency, given in cycles per second or hertz) can be applied to the sample. After an initial start-up period, a stress develops in response to the applied strain, that is, it oscillates with the same frequency. The change of the sine waves of the stress and strain with time can be analyzed to distinguish between elastic, viscous, and viscoelastic response. Analysis of the resulting sine waves can be used to obtain the various viscoelastic parameters as discussed in the following text.

Let us consider the case of a viscoelastic system. The sine waves of strain and stress are shown in Figure 8.15. The frequency ω is in radians per second and the time shift between strain and stress sine waves is Δt. The phase angle shift δ is given by (in dimensionless units of radians)

$$\delta = \omega \Delta t \tag{8.12}$$

As discussed earlier

- perfectly elastic solid $\delta = 0$
- perfectly viscous liquid $\delta = 90°$
- viscoelastic system $0 < \delta < 90°$.

The ratio of the maximum stress σ_o to the maximum strain γ_o gives the complex modulus $|G^*|$

$$|G^*| = \frac{\sigma_o}{\gamma_o} \tag{8.13}$$

$|G^*|$ can be resolved into two components: storage (elastic) modulus G', the real component of the complex modulus and loss (viscous) modulus G'', the imaginary component of the complex modulus.

$$|G^*| = G' + iG'' \tag{8.14}$$

where i is the imaginary number that is equal to $(-1)^{1/2}$.

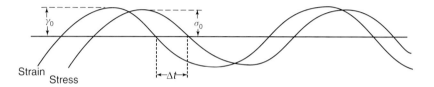

Δt = time shift for sine waves of stress and strain.

$\Delta t \omega = \delta$ Phase angle shift

ω = frequency in rad s^{-1}

$\omega = 2\pi\nu$

Perfectly elastic solid $\delta = 0$
Perfectly viscous liquid $\delta = 90°$
Viscoelastic system $0 < \delta < 90°$

Figure 8.15 Strain and stress sine waves for a viscoelastic system.

The complex modulus can be resolved into G' and G'' using vector analysis and the phase angle shift δ as shown below.

$$G' = |G^*|\cos\delta \tag{8.15}$$

$$G'' = |G^*|\sin\delta \tag{8.16}$$

$$\tan\delta = \frac{G''}{G'} \tag{8.17}$$

Dynamic viscosity η'

$$\eta' = \frac{G''}{\omega} \tag{8.18}$$

Note that $\eta \to \eta(0)$ as $\omega \to 0$.

In oscillatory techniques, two types of experiments can be carried out.

- **Strain sweep**: The frequency ω is kept constant (say at 1 Hz or 6.28 rad s^{-1}) and G^*, G', and G'' are measured as a function of strain amplitude.
- **Frequency sweep**: The strain is kept constant (in the linear viscoelastic region) and G^*, G', and G'' are measured as a function of frequency.

As an illustration, Figure 8.16 shows the strain sweep profile where G^*, G', and G'' remain constant up to a critical strain γ_{cr}. This is the linear viscoelastic region where the moduli are independent of the applied strain. Above γ_{cr}, G^* and G' start

Fixed frequency (0.1 or 1 Hz) and follow G^*, G' and G'' with strain amplitude γ_0

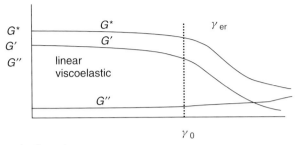

Linear viscoelastic region

G^*, G' and G'' are independent of strain amplitude

γ_{cr} is the critical strain above which system shows nonlinear response (break down of structure)

Figure 8.16 Schematic representation of strain sweep.

to decrease, whereas G'' starts to increase with further increase in γ_0. This is the nonlinear region.

γ_{cr} may be identified with the critical strain above which the structure starts to "break down." It can also be shown that above another critical strain, G'' becomes higher than G'. This is sometimes referred to as the *"melting strain"*, at which the system becomes more viscous than elastic [15].

Figure 8.17 shows the frequency sweep trends for a viscoelastic liquid. In the low-frequency regime, that is, $\omega < \omega^*$, $G'' > G'$. This corresponds to a long time experiment (time is reciprocal of frequency), and hence, the system can dissipate energy as viscous flow. In the high-frequency regime, that is, $\omega > \omega^*$, $G' > G''$. This corresponds to a short time experiment, where energy dissipation is reduced.

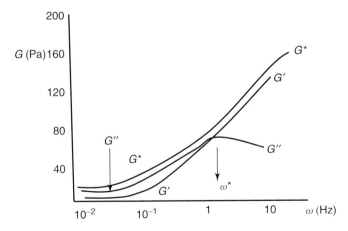

Figure 8.17 Schematic representation of oscillatory measurements for a viscoelastic liquid.

At sufficiently high frequency, $G' \gg G''$, and at such high frequency, $G'' \to 0$ and $G' \sim G^*$. The high-frequency modulus $G'(\infty)$ is sometimes referred to as the *rigidity modulus* where the response is mainly elastic.

8.6
Rheology of Oil-in-Water (O/W) Emulsions Stabilized with Poly(Vinyl Alcohol)

Liquid paraffin-in-water emulsions were prepared using a commercial sample of PVA that has an average molecular weight of 45 000 Da and contains 12% acetate groups. An emulsion containing a volume fraction of oil of 0.6 was prepared using 5% PVA in the aqueous phase. The emulsification was carried out using an UltraTurrax that was run at full speed for 15 min. The droplet size of the emulsion was determined using a Coulter counter. The mean volume radius was found to be 0.94 µm and it did not change after storage for 10 days at room temperature, indicating the absence of coalescence. The rheological measurements were carried out using a Bohlin rheometer (Bohlin Instruments, UK). A coaxial cylinder with a moving cup of 27.5 mm and a fixed bob of radius 25.0 mm was used. The Bohlin rheometer performs steady shear viscosity measurements by incrementing or decrementing the shear rate in discrete steps. At each shear rate, 10 s is allowed for the linear stress to reach a steady state value and then the average shear stress is measured over 3 s. Each sample was sheared at 14.7 s^{-1} for 300 s and rested for 600 s before measurement. The viscosities are quoted for the decreasing shear rate cycle of an up–down plot. The yield stress is obtained by extrapolation of the linear region of a shear stress–shear rate curve to zero shear rate (using the Bingham model). The Bohlin rheometer performs oscillatory measurements by turning the cup back and forth, in a sinusoidal movement. The shear stress in the sample is measured by measuring the deflection of the bob, which is connected to interchangeable torsion bars. A torsion bar with a constant of 3.73 g cm was used. The movement is detected by the movement of a magnetically permeable core in a fixed coil.

8.6.1
Effect of Oil Volume Fraction on the Rheology of the Emulsions

Shear stress–shear rate curves were obtained at various oil volume fractions: $\phi = 0.40, 0.43, 0.47, 0.50, 0.53, 0.55, 0.57, 0.59$, and 0.60. When $\phi < 0.45$, the flow curves were nearly Newtonian. At $\phi > 0.45$, the emulsions showed non-Newtonian flow (shear thinning) with a measurable extrapolated yield value (using Bingham model). This reflects the interaction between the adsorbed PVA layers with increase in ϕ. When $\phi < 0.45$, the surface-to-surface distance h between the emulsion droplets is much higher than twice the adsorbed layer thickness δ of PVA. When $\phi > 0.5$, h starts to become comparable to 2δ and the adsorbed PVA "tails" begin to interact. The exact oil volume fraction at which significant interaction occurs can

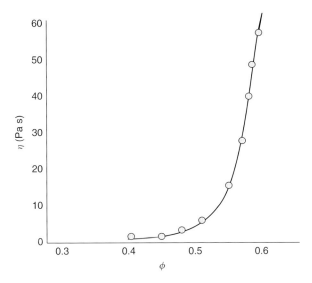

Figure 8.18 Variation of viscosity of PVA-stabilized O/W emulsions with the oil volume fraction.

be obtained from plots of viscosity η and yield value σ_β versus ϕ as is illustrated in Figures 8.18 and 8.19.

It can be seen from Figures 8.18 and 8.19 that both the viscosity and yield value show a rapid increase, when the oil volume fraction is >0.52. Assuming the droplets to behave as hard spheres, this rapid increase should occur when ϕ approaches maximum packing. The latter (maximum packing fraction) for monodisperse

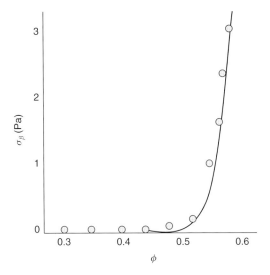

Figure 8.19 Variation of yield value of PVA-stabilized O/W emulsion with the oil volume fraction.

systems is in the region of 0.6 (assuming random packing). This means that the effective volume fraction at this rapid increase is in the region of 0.6 and this can be used to calculate the adsorbed layer thickness Δ of PVA using the following equation:

$$\phi_{\text{eff}} = \phi \left[1 + \frac{\Delta}{R}\right]^3 \tag{8.19}$$

Using Eq. (8.20) a value of $\Delta \sim 50$ nm is obtained, which is significantly higher than the value obtained on polystyrene latex ($\sim$26 nm), using dynamic light scattering. This disagreement is not surprising for two main reasons: Firstly, the emulsion droplets do not behave as hard spheres; secondly, the emulsion is polydisperse and using the average radius is only approximate. However, the results obtained showed that the adsorbed layer thickness of PVA is quite high because of the presence of long dangling tails. Further evidence for the interaction between PVA-stabilized emulsions was obtained using viscoelastic measurements as a function of the oil volume fraction. As discussed above, the moduli were measured as a function of strain amplitude to obtain the linear viscoelastic region. Measurements were then carried out as a function of frequency (0.01–5 Hz), while keeping the strain constant in the linear region. The results clearly showed a transition from predominantly viscous ($G'' > G'$) to predominantly elastic ($G' > G''$) response, as the oil volume fraction was increased.

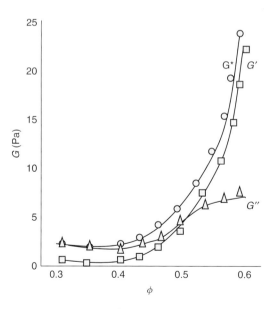

Figure 8.20 Variation of G^*, G', and G'' with oil volume fraction for PVA-stabilized O/W emulsion.

This is illustrated in Figure 8.20, which shows a plot of G^*, G', and G'' versus ϕ at a frequency of 1 Hz and strain of 0.01 (in the linear viscoelastic region).

When $\phi < 0.52$, $G'' > G'$ indicating the weak interaction between the adsorbed PVA layers, as in this case $h > 2\delta$. At $\phi = 0.52$, $G' = G''$ (tan $\delta = 1$) and this denotes the onset of strong steric interaction. At $\phi > 0.52$, $G' > G''$ and the magnitude of G' increases rapidly with further increase in ϕ, whereas G'' (which is lower than G') shows only a slow increase with increase in ϕ. Indeed, at sufficiently high ϕ approaching maximum packing (~ 0.6), G' approaches G^* ($G' = 22$ Pa, whereas $G^* = 23$ Pa), whereas G'' remains low (in the region of ~ 7 Pa). This behavior reflects the strong steric repulsion between the PVA-stabilized emulsions. When the volume fraction of the oil increases such that the interdroplet distance becomes smaller than twice the adsorbed layer thickness, interpenetration and possible compression of the long dangling tails of PVA occurs, resulting in strong elastic response. Evidence for this strong steric repulsion between PVA layers was obtained by Sonntag et al., [18] who measured the interaction forces between two crossed quartz filaments both containing an adsorbed PVA layer (with $M = 55\,000$ and containing 12% acetate groups). The interaction force increased sharply with decrease of separation distance h, when the latter was <110 nm indicating an adsorbed layer thickness of 55 nm, which is higher than our result of 50 nm, as the molecular weight of PVA used in Sonntag et al. experiment is higher.

8.6.2
Stability of PVA-Stabilized Emulsions

The stability of the emulsions was investigated by following the effect of addition of electrolyte (KCl or Na_2SO_4) on the viscosity, yield value, and viscoelastic parameters, at constant volume fraction of the emulsion ($\phi = 0.535$), and at constant temperature. Alternatively, the stability was assessed by measuring the rheological parameters as a function of temperature, at constant electrolyte concentration.

Figure 8.21 shows the variation of viscosity with KCl concentration, and Figure 8.22 shows the corresponding variation of yield stress.

It can be seen from Figures 8.21 and 8.22 that both η and σ_β show a small decrease in their values with increase in KCl concentration up to 1.5 mol dm^{-3}. This can be attributed to the reduction of solvency of the PVA chains on addition of electrolyte and hence a reduction in the adsorbed layer thickness. This decrease results in a reduction of the effective volume fraction of the emulsion and this will be accompanied by reduction in η and σ_β. However, above 1.5 mol dm^{-3} KCl, there is a sharp increase in η and σ_β. Under these conditions, the continuous medium becomes worse than a θ-solvent, resulting in incipient flocculation. This indicates that 1.5 mol dm^{-3} KCl is the critical flocculation concentration (CFC) of this electrolyte, at constant temperature.

A more sensitive method to determine the CFC is dynamic or oscillatory measurements. This is illustrated in Figure 8.23, which shows the variation of G'

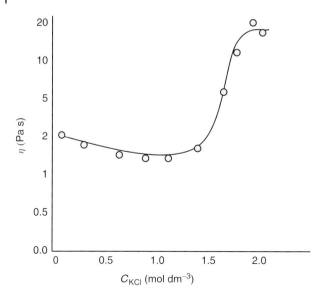

Figure 8.21 Variation of viscosity with KCl concentration.

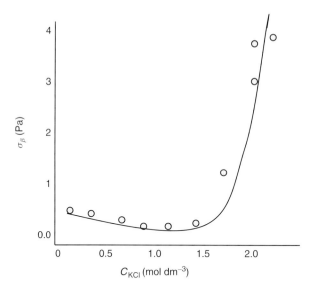

Figure 8.22 Variation of yield value with KCl concentration.

and G'' (at an amplitude of 0.01 in the linear viscoelastic region and frequency of 1 Hz) with KCl concentration.

It can be seen from Figure 8.23 that both G' and G'' initially decrease with increase in KCl concentration due to the reduction of the effective volume fraction of the emulsion (as a result of reduction of the adsorbed layer thickness). At

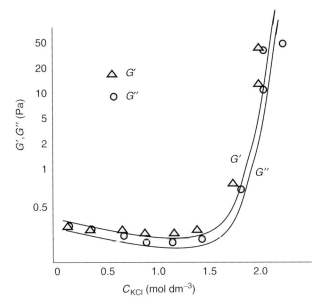

Figure 8.23 Variation of G' and G'' with KCl concentration.

$C_{KCl} > 1.5$ mol dm^{-3} both G' and G'' show a rapid increase with further increase in KCl concentration.

Similar trends were obtained on addition of Na$_2$SO$_4$ as is illustrated in Figures 8.24–8.26, which show the variation of viscosity, yield value, and G' and G'' with increase in Na$_2$SO$_4$ concentration. In this case, the CFC is ~0.18 mol dm^{-3}.

As mentioned above, the stability of the emulsion was assessed by measuring the critical flocculation temperature (CFT), whereby the rheological parameters were determined as a function of temperature, at constant electrolyte concentration. By increasing the temperature, the hydration of the PVA chains decreases, and at the CFT the emulsion reaches its θ-condition. This results in catastrophic flocculation with a rapid increase in the viscosity, the yield value, and the elastic modulus. This is illustrated in Figures 8.27–8.29 for an emulsion at 1.18 mol dm^{-3}, whereby η, σ_β, G', and G'' are measured, as a function of temperature in the range 15–40 °C. The viscosity (Figure 8.27) remains virtually constant with increase of temperature in the range 15–28 °C, indicating a stable emulsion in this temperature range. Above 28 °C, the viscosity begins to increase and a rapid increase is observed above 30 °C. These results indicate a CFT of ~30 °C. A similar trend is obtained from plot of σ_β versus temperature (Figure 8.28). However, the results of the moduli measurements show an initial reduction in G' and G'' with increase in temperature in the range 15–32 °C that is consistent with the reduction in the effective volume fraction of the emulsion as a result of the decrease of the adsorbed layer thickness with increase of temperature (due to dehydration of the PVA chains). The sharp increase in G' and G'' occurs above 32 °C that is considered the CFT of the emulsion. The small difference in CFT obtained from viscosity and yield value (~30 °C) and

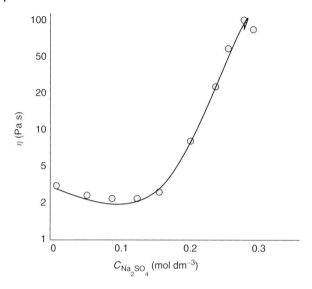

Figure 8.24 Variation of viscosity with Na_2SO_4 concentration.

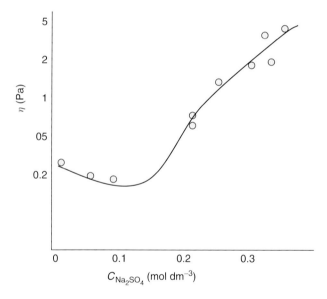

Figure 8.25 Variation of yield value with Na_2SO_4 concentration.

the moduli measurements (32 °C) is not surprising, as the results obtained from viscosity and yield value are subject to some uncertainty, as the system is subjected to high deformation in the flow curve measurements.

Similar results were obtained when using emulsions in the presence of 0.175 mol dm^{-3} Na_2SO_4 as is illustrated in Figures 8.30–8.32 that show the

8.6 Rheology of Oil-in-Water (O/W) Emulsions Stabilized with Poly(Vinyl Alcohol) | 233

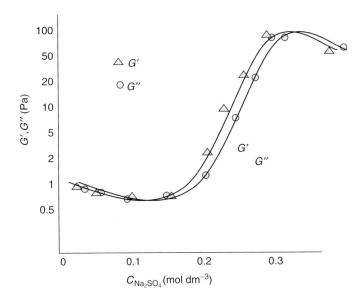

Figure 8.26 Variation of G' and G'' with Na_2SO_4 concentration.

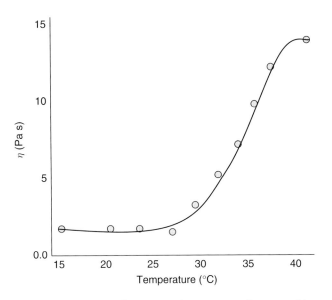

Figure 8.27 Variation of viscosity with temperature for an emulsion stabilized with PVA at 1.18 mol dm^{-3} KCl.

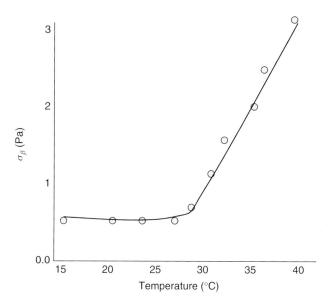

Figure 8.28 Variation of yield value with temperature for an emulsion stabilized with PVA at 1.18 mol dm^{-3} KCl.

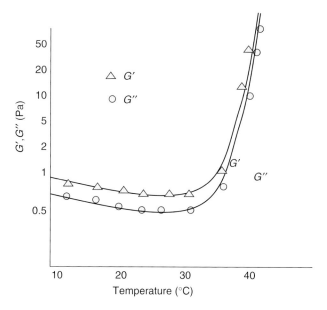

Figure 8.29 Variation of G' and G'' with temperature for an emulsion stabilized with PVA at 1.18 mol dm^{-3} KCl.

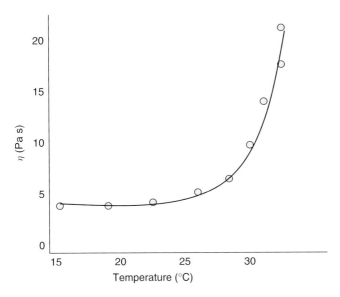

Figure 8.30 Variation of viscosity with temperature for an emulsion stabilized with PVA at 0.175 mol dm^{-3} Na$_2$SO$_4$.

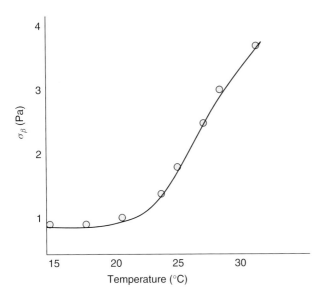

Figure 8.31 Variation of yield value with temperature for an emulsion stabilized with PVA at 0.175 mol dm^{-3} Na$_2$SO$_4$.

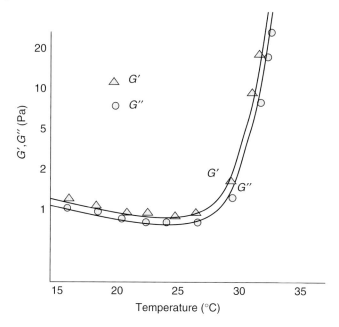

Figure 8.32 Variation of G' and G'' with temperature for an emulsion stabilized with PVA at 0.175 mol dm^{-3} Na$_2$SO$_4$.

variation of viscosity, yield value, and G' and G'' with temperature. The CFT value obtained using viscosity and yield value measurements is $\sim$22 °C, whereas the value obtained using oscillatory measurements is 27 °C. The latter value is more accurate as measurements of G' and G'' were obtained at low amplitude and high frequency, in which case the structure of the flocculated emulsion is not significantly deformed.

8.6.3
Emulsions Stabilized with an A-B-A Block Copolymer

The emulsions were prepared using Isopar M (medium chain paraffinic oil, supplied by Exxon) and Pluronic L92 (an A-B-A block copolymer of PEO-PPO-PEO) with 47.3 PO units and 15.6 EO units). A stock emulsion with a volume fraction ϕ of 0.6 was prepared using 15% Pluronic L92 based on the oil phase. The emulsification was carried out using an Ultra Turrax at 20 000 rpm. The droplet size distribution of the emulsion was determined using a Coulter counter. The results are shown in Figure 8.33, which gives the histograms of the percentage volume versus diameter in microns.

The results of Figure 8.33 show that the emulsion is fairly polydisperse giving a mean volume diameter of 098 μm. The stock emulsion was diluted with water to give oil volume fractions in the range 0.48–0.60. Oscillatory measurements were carried out for each emulsion as a function of strain amplitude to obtain the linear

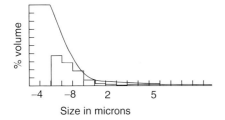

Figure 8.33 Droplet size distribution of Isopar oil/water emulsions stabilized by Pluronic L92.

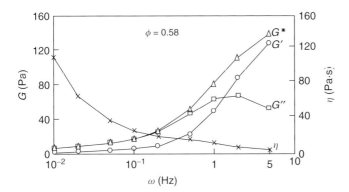

Figure 8.34 Variation of G^*, G', G'', and η with frequency at $\phi = 0.58$.

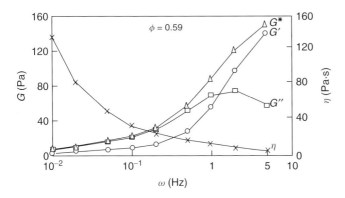

Figure 8.35 Variation of G^*, G', G'', and η with frequency at $\phi = 0.59$.

viscoelastic region. Measurements were then carried out as a function of frequency, while keeping the strain amplitude constant (in the linear region).

Figures 8.34–8.36 show the variation of G^*, G', G'', and dynamic viscosity as a function of frequency ω (Hz). These results show the typical behavior of a viscoelastic liquid (that is represented by a Maxwell model), where G' increases with increases of frequency and approaches G^*, at high frequency (5 Hz). G'' increases with increase of frequency, reaches a maximum at a frequency of 1–2 Hz

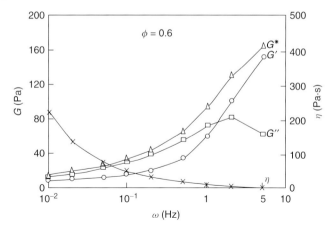

Figure 8.36 Variation of G^*, G', G'', and η with frequency at $\phi = 0.6$.

and then decreases with further increase of frequency. In the low-frequency regime (long time) $G'' > G'$, whereas in the high-frequency regime $G' > G''$. At a characteristic frequency ω^* (the crossover point), $G' = G''$ ($\tan \delta = 1$, where δ is the phase angle shift). From ω^* one can calculate the Maxwell relaxation time t^*,

$$t^* = \frac{1}{\omega^*} \tag{8.20}$$

t^* is plotted as a function of the oil volume fraction of the emulsion as illustrated in Figure 8.37. The results show a great deal of scatter at volume fractions below

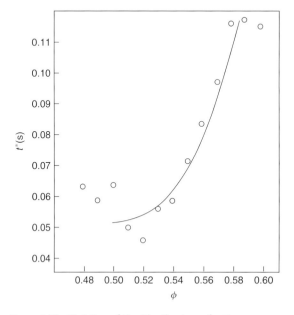

Figure 8.37 Variation of t^* with oil volume fraction.

0.565. This may be due to the inaccuracy of determination of the crossover point below this volume fraction. However, above this value, there is a sharp increase in t^* with increase in ϕ. This increase in Maxwell relaxation time with increase in ϕ is due to the increase in interaction between the sterically stabilized emulsion as the droplet come close to each other. A useful method for determination of the onset of strong steric interaction is obtained from plots of G^*, G', and G'' (at a frequency of 1 Hz) versus the oil volume fraction. The results are shown in Figure 8.38. When ϕ is <0.565, $G'' > G'$. Under these conditions, the surface-to-surface distance h between the emulsion droplets is smaller than twice the adsorbed PEO layers 2δ of the block copolymer. At $\phi = 0.565$, the PEO layers begin to overlap and the steric interaction resulting from these layers increases with decrease of h.

A rough estimate of the adsorbed layer thickness may be obtained from the volume fraction of the oil at the crossover point at which $G' = G''$. Assuming the packing of the oil droplets is random in nature at this point, one can assume that the effective volume fraction ϕ_{eff} at the crossover point is ~ 0.6. From ϕ_{eff} and the

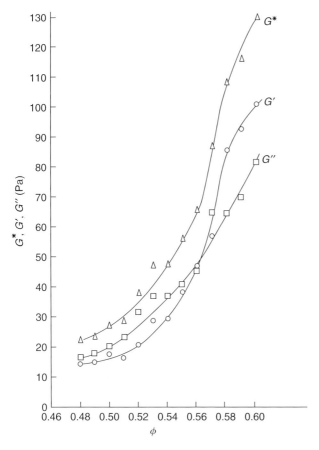

Figure 8.38 Variation of G^*, G', and G'' with core oil volume fraction.

core volume fraction ϕ, the adsorbed layer thickness δ can be calculated using Eq. (8.20). This gives a value of δ in the region of ~15 nm, which is an overestimate for the PEO chain of the block copolymer (~8 EO units). However, it should be mentioned that the emulsion is polydisperse and using the average droplet radius R for estimation of δ can lead to significant errors.

8.6.4
Water-in-Oil Emulsions Stabilized with A-B-A Block Copolymer

A W/O emulsion was prepared at a volume fraction ϕ of 0.75 by emulsification of water into an Isopar M oil solution containing an A-B-A block copolymer of PHS-PEO-PHS. The polymer was dissolved in the oil by warming using a hot plate. The oil phase consisted of 59.5 ml of PHS-PEO-PHS and 190.5 ml of Isopar M, whereas the emulsified water phase was 750 ml. The emulsification was carried out using an Ultra Turrax at 10 000 rpm for 5 min, while cooling the system using an ice bath. The droplet size distribution of the emulsion was determined using dynamic light scattering (photon correlation spectroscopy, PCS) using a Malvern instrument (Malvern, UK). The emulsion was extensively diluted with Isopar M to avoid multiple scattering. The emulsion has a narrow droplet size distribution with a z-average radius of 183 nm. The stock emulsion was then diluted with Isopar M to obtain systems covering the volume fraction range 0.4–075. Figure 8.39 shows the shear stress τ–shear rate $\dot{\gamma}$ curves for the emulsions studied. It can be seen that when $\phi < 0.5$, the emulsion shows Newtonian flow, whereas when $\phi > 0.5$ non-Newtonian (pseudoplastic) flow is observed. The non-Newtonian flow becomes more pronounced as ϕ increases.

The results of Figure 8.39 were analyzed using the Bingham equation (Eq. (8.6)) to obtain the variation of yield value and plastic viscosity with volume fraction. The results are shown in Figure 8.40.

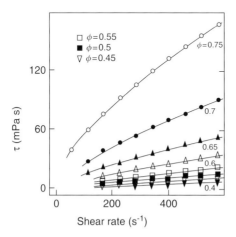

Figure 8.39 Shear stress–shear rate curves for W/O emulsions at various volume fractions.

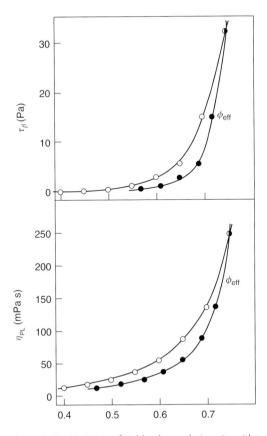

Figure 8.40 Variation of yield value and viscosity with emulsion volume fraction.

The viscosity–volume fraction relationship may be used to obtain the adsorbed layer thickness as a function of the droplet volume fraction ϕ. Assuming the W/O emulsion behaves as a near hard-sphere dispersion, it is possible to apply the Dougherty–Krieger equation (8.22) to obtain the effective volume fraction ϕ_{eff}. The assumption that the W/O emulsion droplets behave as near hard spheres is reasonable, as the water droplets are stabilized with a block copolymer with a relatively short PHS chains (of the order of 10 nm or less). Any lateral displacement will be opposed by the high Gibbs elasticity of the adsorbed layer [5] and the droplets will maintain their spherical shape under the conditions of the rheological experiments.

For hard-sphere dispersions, the relative viscosity η_r is related to the volume fraction ϕ by the following equation:

$$\eta_r = \left[1 - \frac{\phi}{\phi_p}\right]^{-[\eta]\phi_p} \tag{8.21}$$

In Eq. (8.22), ϕ is replaced by ϕ_{eff} that includes the contribution from the adsorbed layer. $[\eta]$ is the intrinsic viscosity, which for hard spheres is equal to 2.5. ϕ_p is the maximum packing fraction. It has been shown [19] that a plot of $1/\eta_r^{1/2}$ versus ϕ is linear with an intercept that is equal to ϕ_p. For the present W/O emulsion, such a plot gave a ϕ_p value of 0.84. This is higher than the maximum packing fraction for uniform spheres (0.74 for hexagonal packing). However, this high value is reasonable considering the polydispersity of the emulsion.

Using ϕ_p and the measured values of η_r, ϕ_{eff} was calculated at each ϕ value, using a theoretical $\eta_r - \phi$ curve and applying Eq. (8.22). From ϕ_{eff} the adsorbed layer thickness Δ was calculated using Eq. (8.20). A plot of Δ versus ϕ is shown in Figure 8.41. It can be seen that Δ decreases with increase of ϕ. The value at $\phi = 0.4$ is $\sim$10 nm, which is a measure of the fully extended PHS chain. At such relatively low ϕ value, there will be no interpretation or compression of the PHS chains, as the surface-to-surface distance between the droplets is relatively large. This value is in close agreement with the results obtained from thin film results using direct measurement between two water droplets [20]. It is also in agreement with the results obtained by Ottewill and coworkers [21] using compression and small-angle neutron scattering.

The decrease of Δ with increase in ϕ is similar to our results [22] for latex dispersions stabilized with grafted PEO chains. This reduction in Δ with increase in ϕ can be attributed to the interpenetration and/or compression of the chains on increasing ϕ. It is likely that the adsorbed PHS layer is not dense enough, and hence, the PHS chains can interpenetrate on close approach of the droplets. If complete interpenetration is possible, the value of Δ can be reduced by half of that obtained in dilute emulsion. Indeed, the results of Figure 8.41 show that Δ is reduced to $\sim$4 nm at $\phi = 0.65$ and this value is close to the predicted value of $\sim$5 nm, considering the errors in the measurements. The reduction in Δ can also be attributed to compression of the chains, without need of invoking

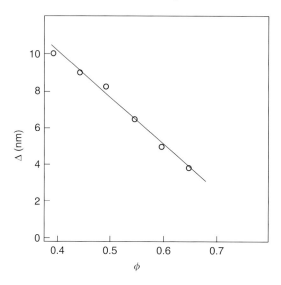

Figure 8.41 Variation of Δ with ϕ.

any interpenetration. The most likely picture is probably a combination of both interpenetration and compression as mentioned above.

The increase in interaction between the droplets with increase in ϕ is also shown by oscillatory measurements. This is illustrated in Figure 8.42, which shows plots of G' and G'' as a function of frequency v (Hz) for three ϕ values of 0.6, 0.675, and 0.75. At $\phi = 0.6$, $G'' > G'$ over the whole frequency range studied. At $\phi = 0.675$, $G'' > G'$ below a critical frequency, above which $G' > G''$. The frequency at which $G' = G''$ (the crossover point) is the characteristic frequency v^*, which is related to the Maxwell relaxation time t^* by

$$t^* = \frac{1}{2\pi v^*} \tag{8.22}$$

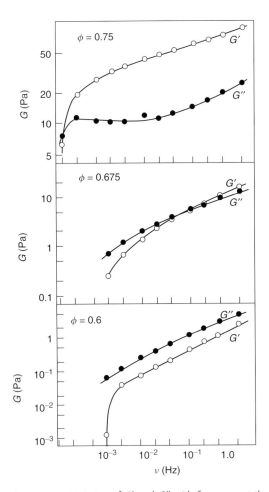

Figure 8.42 Variation of G' and G'' with frequency at three emulsion volume fractions.

As ϕ is increased above 0.675, ν^* shifts to lower frequency, that is, longer Maxwell relaxation time. A plot of G' and G'' (at $\nu = 1$ Hz) versus ϕ is shown in Figure 8.43, which illustrates the critical volume fraction ϕ_* above which the emulsion becomes more elastic than viscous. The same figure also shows the variation of G'/G'' versus ϕ. It can be clearly shown that when $\phi > 0.675$, G' and G'/G'' increases very rapidly with increase in ϕ indicating the strong elastic interaction between the PHS chains as the surface-to-surface distance between the droplets becomes smaller than 2Δ. This strong elastic interaction is also reflected in the rapid increase of the Maxwell relaxation time with increase in ϕ as is illustrated in Figure 8.44.

The viscosity data showed interpenetration of the chains with increase in ϕ. However, such interpenetration has to become significant before an elastic response is observed. This seems to occur at a ϕ value of 0.67, where there is probably full

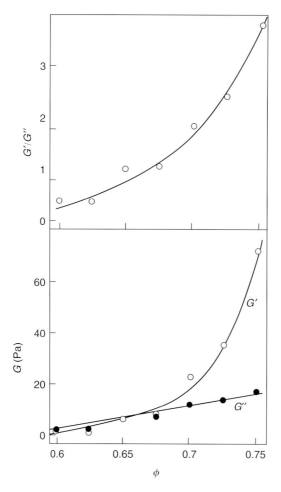

Figure 8.43 Variation of G', G'', and G'/G'' with ϕ.

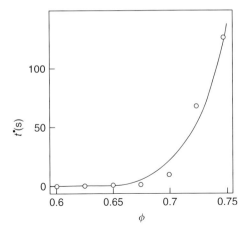

Figure 8.44 Variation of Maxwell relaxation time t^* with emulsion volume fraction.

interpenetration of the chains. Above $\phi = 0.67$, compression of the chains occurs and this leads to a more elastic response, which increases with increase in ϕ as expected (Figure 8.43). This elastic response is also reflected in the increase in relaxation time t^*, as shown in Figure 8.44. The latter shows an initial slow increase in t^* with increase in ϕ, but above ϕ_c, that is 0.67, t^* increases rapidly with increase in ϕ. This increase in relaxation time reflects the highly elastic nature of the emulsion structure, in which the droplets probably form a three-dimensional arrangement with the PHS chains strongly repelling each other as a result of the good solvency of the medium for the chains.

References

1. Tadros, T.F. (2005) *Applied Surfactants*, Wiley-VCH Verlag GmbH, Germany.
2. Tadros, Th.F. (1999) in *Principles of Polymer Science and Technology in Cosmetics and Personal Care* (eds E.D. Goddard and J.V. Gruber), Marcel Dekker, New York.
3. Tadros, T. (2003) in *Novel Surfactants* (ed. K. Holmberg), Marcel Dekker, New York.
4. Tadros, Th.F. and Vincent, B. (1983) in *Encyclopedia of Emulsion Science and Technology*, vol. 1 (ed. P. Becher), Marcell Dekker, New York.
5. Binks, B.P. (ed.) (1998) *Modern Aspects of Emulsion Science*, Royal Society of Chemistry Publication, Cambridge.
6. Piirma, I. (1992) *Polymeric Surfactants*, Surfactant Science Series 42, Marcel Dekker, New York.
7. Stevens, C.V., Meriggi, A., Peristerpoulou, M., Christov, P.P., Booten, K., Levecke, B., Vandamme, A., Pittevils, N., and Tadros, Th.F. (2001) *Biomacromolecules*, **2**, 1256.
8. Hirst, E.L., McGilvary, D.I., and Percival, E.G. (1950) *J. Chem. Soc.*, 1297.
9. Suzuki, M. (1993) in *Science and Technology of Fructans* (eds M. Suzuki and N.J. Chatterton), CRC Press, Boca Raton, FL, p. 21.
10. Napper, D.H. (1983) *Polymeric Stabilization of Dispersions*, Academic Press, London.
11. Flory, P.J. and Krigbaum, W.R. (1950) *J. Chem. Phys.*, **18**, 1086.
12. Tadros, Th.F., Vandamme, A., Levecke, B., Booten, K., and Stevens, C.V. (2004) *Adv. Colloid Interface Sci.*, **108–109**, 207.

13. Exerowa, D., Gotchev, G., Kolarev, T., Khristov, K., Levecke, B., and Tadros, T. (2007) *Langmuir*, **23**, 1711.
14. Krieger, I.M. (1972) *Adv. Colloid Interface Sci.*, **3**, 111.
15. Tadros, T. (2010) *Rheology of Dispersions*, Wiley-VCH Verlag GmbH, Germany.
16. Bingham, E.C. (1922) *Fluidity and Plasticity*, McGraw Hill, New York.
17. (a) Herschel, W.H. and Bulkley, R. (1926) *Proc. Am. Soc. Test Mater.*, **26**, 621; (b) Herschel, W.H. and Bulkley, R. (1926) *Kolloid-Z.*, **39**, 291.
18. Sonntag, H. Ehmke, B. Miller, R. and Knapschinsky, L. (1981), in *The effect of Polymers on Dispersion Properties.*, Th.F. Tadros(Editor), Academic Press, pp 207–220.
19. Prestidge, C. and Tadros, Th.F. (1988) *Colloids Surf.*, **31**, 325.
20. Aston, M.S., Herrington, Th.M., and Tadros, Th.F. (1989) *Colloids Surf.*, **40**, 49.
21. Ottewill, R.H. (1987) in *Solid/Liquid Dispersions* (ed. Th.F. Tadros), Academic Press, London.
22. Liang, W., Tadros, Th.F., and Luckham, P.F. (1992) *J. Colloid Interface Sci.*, **153**, 131.

Index

a

A-B-A block copolymer
– emulsions stabilized with 236–240
– water-in-oil emulsions stabilized with 240–245
adsorption 9, *10*, *12*
– characteristics, of ionic surfactants 99
– – experimental tools 99–101
– – results 102–107
– – theory 101–102
– of surfactants at liquid/liquid interface 14
– – emulsification mechanism 17–19
– – emulsification methods 19–20
– – Gibbs adsorption isotherm 14–17
– – role in droplet formation 22–26
– – role in emulsion formation 21–22
anchoring chain 209
anisotropic etching 86
anisotropy 66
average energy dissipation rate 144

b

bagasse particleboard 170
Bancroft rule 24
batch mixers *129*, 132–133, *134*
batch rotor–stator mixers 154–157
Bell-shaped conductivity curves 203
Bingham plastic systems 221
bitumen emulsions 5
Bohlin rheometer (Bohlin Instruments, UK) 226
breakup conditions 69
Brownian diffusion 35, 37
Brownian motion 3, 54, 193
bulk rheology of emulsions 53–54
– concentrated emulsions behavior analysis 54–57

c

calibration curves 119
capillary method 118
capillary number 140, 141
capillary pressure 66, 81
catastrophic inversion 47
cationic emulsifiers 81
characteristic length 44
charge-stabilized emulsions 43
classical model, of particle size 182, 183
coalescence 4, 45–46, 161, 185
– phase inversion 47–48
– rate 46–47
cohesive energy ratio (CER) 31–32
cohesive stress 140
colloid mills 129–130
comb stabilizer 211
concentrated emulsions viscoelastic properties 59–60
– deformation and droplet breakup in emulsions during flow 66–72
– high interval phase emulsions (HIPEs) 61–65
constant stress (creep) measurements 222–223
Coulter counter 226
counteracting stress 24
creaming and sedimentation 3, 35–36
– prevention 37–40
– rates 36–37
creep measurements, *see* constant stress (creep) measurements
critical aggregation concentration (CAC) 182–183
critical capillary number 141
critical coagulation concentration (CCC) 189, 190, 194

Emulsion Formation and Stability, First Edition. Edited by Tharwat F. Tadros.
© 2013 Wiley-VCH Verlag GmbH & Co. KGaA. Published 2013 by Wiley-VCH Verlag GmbH & Co. KGaA.

critical flocculation concentration (CFC) 42, 43, 229
critical flocculation temperature (CFT) 231
critical micelle concentration (CMC) 103
critical packing parameter (CPP) for emulsion selection 32–35
critical volume fraction (CFV) 43
curved interface 6

d

Debye–Hückel equation 102
Debye–Hückel parameter 191–192
deep reactive ion etching (DRIE) 88
deformation and droplet breakup in emulsions during flow 66–72
Derjaguin approximation 191
differential scanning calorimetry 170
Difftrain 119
dilational elasticity, interfacial 50
dilational viscoelasticity 100, *105*, 106
dilational viscosity, interfacial 51
direct imaging 115–118
direct membrane emulsification 78–79
disjoining pressure 45
disruptive stress 140
DLVO theory 11, 190, 193, 194
double and multiple emulsions 2
double-layer extension 10–11
Dougherty–Krieger equation 219, 241
droplet coalescence investigation techniques 121–123
droplet concentration 36
droplet disruption 173
droplet size reduction 37
drop profile tensiometry 100
drop size distributions
– and average drop sizes 138–140
– prediction, during emulsification 160–163
dynamic (oscillatory) measurements 223–226

e

Einstein equation 53
elastic interaction 213, 214–216
electrolytes, paraffin emulsion stability as function of 189–195
electrophoretic mobility measurements 189, 192
electrostatic potential 191
electrostatic repulsion 9–12
electrosteric stabilization 175, 195
elongational flow 141
empirical models 161
emulsions 1. See also individual entries
– adsorption of surfactants at liquid/liquid interface 14
– – emulsification mechanism 17–19
– – emulsification methods 19–20
– – Gibbs adsorption isotherm 14–17
– – role in droplet formation 22–26
– – role in emulsion formation 21–22
– breakdowm processes 3
– bulk rheology of emulsions 53–54
– – concentrated emulsions behavior analysis 54–57
– coalescence 4, 45–46
– – phase inversion 47–48
– – rate 46–47
– concentrated emulsions viscoelastic properties 59–60
– – deformation and droplet breakup in emulsions during flow 66–72
– – high interval phase emulsions (HIPEs) 61–65
– creaming and sedimentation 3, 35–36
– – prevention 37–40
– – rates 36–37
– emulsifier nature 1–2
– emulsifiers selection 26
– – cohesive energy ratio (CER) 31–32
– – critical packing parameter (CPP) for emulsion selection 32–35
– – hydrophilic–lipophilic balance (HLB) 26–29
– – phase inversion temperature (PIT) 29–31
– experimental η_r-φ curves 57–58
– – droplet deformability influence 58–59
– flocculation 4
– – mechanism 40–43
– – rules for reducing 43–44
– industrial applications 4–5
– interaction energies between emulsion droplets and combinations
– – electrostatic repulsion 9–12
– – steric repulsion 12–13
– – van der Waals attraction 8–9
– Ostwald ripening 4, 44–45
– phase inversion 4
– physical chemistry of systems
– – interface (Gibbs dividing line) 5–6
– rheology 48
– – and emulsion stability correlation 51–53
– – interfacial dilational elasticity 50
– – interfacial dilational viscosity 51
– – interfacial shear viscosity measurement 49–50
– – interfacial 48–49

– – non-Newtonian effects   51
– stability correlation   51–53
– system structure   2
– thermodynamics of formation and breakdown   6–8
equation of state approach   14
ethyl acetate   202
ethylcellulose nanoemulsions, for nanoparticle preparation   202–204
experimental η_r–φ curves   57–58
– droplet deformability influence   58–59

f

fiber-optical spot scanning (FSS)   113
flocculation   4, 40, 186
– controlled   39
– depletion   39
– of electrostatically stabilized emulsions   41–42
– mechanism   40–43
– rules for reducing   43–44
– of sterically stabilized emulsions   42–43
Flory–Huggins interaction parameter   213, 214
focused beam reflectance measurement (FBRM)   113
food emulsion   4
Fraunhofer diffraction theory   112
Freeze–Thaw cycles, paraffin emulsion stability as function of   186–189
frequency sweep   224
Frumkin ionic compressibility (FIC)   101, 103–104

g

geometrically mediated breakup   82
Gibbs adsorption equations   15, 183
Gibbs adsorption isotherm   14–17
Gibbs approach   14
Gibbs–Deuhem equation   6, 14
Gibbs dividing line   5–6
Gibbs-Marangoni effect   24–25
glass capillary microfluidic devices   89–93
Grace curves, see stability curves
graft copolymers   211
grafting onto technique   211
Grahame's equation   191
grooved-type microchannel arrays   86–88

h

Hamaker constant   9, 40, 190, 191
Herschel–Bulkley general model   222
high interval phase emulsions (HIPEs)   61–65

high-shear mixers   127, 138
homopolymer   210
Hough transformation   116
hyberbolic flow   141
hydrodynamic diameter   203
hydrodynamic flow focusing   83
hydrophilic–lipophilic balance (HLB)   26–29

i

incipient flocculation   42
industrial applications   4–5
industrial-scale rotor–stator mixers   *134*
in-line mixers   *129*, 130–131, 136–137, 145, 146, 147, 148, 157–160
– radial discharge mixers   *129*, 130–131
interaction energies, between emulsion droplets and combinations
– electrostatic repulsion   9–12
– steric repulsion   12–13
– van der Waals attraction   8–9
interfacial rheology
– emulsion stability correlation   51–53
– interfacial dilational elasticity   50
– interfacial dilational viscosity   51
– interfacial shear viscosity measurement   49–50
– interfacial tension and surface pressure   48–49
– non-Newtonian effects   51
interfacial tension   29
– gradients   23
interfacial tension gradient   50
inulin   211
INUTEC® N25   211
INUTEC® SP1   45, 211, 216
ionic surfactants   17, 43

k

Kolmogorov length scale   142, 143, 150–151
Krafft point   173, 175, 176, 183

l

laminar flow, maximum stable drop size in   141–142
Laplace pressure   6, 17, *18*, 66, 173
Lasentech (USA)   117
laser beam diffraction   112
laser Doppler anemometry (LDA)   145
laser systems   112–115
light transmission method   119
liquid–liquid interface   182, 184
liquid–liquid systems, drop size in   145–147
– two phase   140
low energy emulsification   200

m

Marangoni effect 23
mean drop size 139
measurement techniques 109–112, 118–120
– droplet coalescence investigation techniques 121–123
– online droplet size measurement techniques
– – direct imaging 115–118
– – laser systems 112–115
– – sound systems 115
melting strain 63, 225
membrane and microfluidic devices 77
– droplet application 93
– glass capillary microfluidic devices 89–93
– membrane emulsification 78
– – direct 78–79
– – operating parameters 80
– – premix 79–80
– – surfactants 80–81
– – transmembrane pressure and wall shear stress 81
– microfluidic devices with parallel microchannel arrays 85–86
– – grooved-type 86–88
– – straight-through 88–89
– microfluidic flow-focusing devices (MFFD) 83–84
– microfluidic junctions 82–83
micellar emulsions and microemulsions 2
microfluidic flow-focusing devices (MFFD) 83–84
micronozzle array, straight-through 89
miniemulsion polymerization 201
mixed emulsions 2
mixed surfactant films 46, 51
mixed surfactant system 173, 177, 178, 184
mixing interaction 213–214
Monte Carlo simulation 122
multipass processing, of rotor–stator mixers 136, 137

n

Nakajima model, of particle size 182, 183
nanoemulsions 2, 181, 199
– component choice aspects 201–202
– ethylcellulose, for nanoparticle preparation 202–204
– phase inversion emulsification methods 200
Newton Black Film (NBF) 218–219
nonionic surfactants 17, 174, 175
non-Newtonian effects 51
non-Newtonian flow 220, 221, 226, 240
nuclear magnetic resonance (NMR) 119
numerical simulation, in rotor–stator mixers 154, 162

o

O/S ratio 200, 202, 203
oil-in-water (O/W) 70, 203
– emulsion rheology stabilized with poly(vinyl alcohol) 226
– – emulsions stabilized with A-B-A block copolymer 236–240
– – oil volume fraction effect 226–229
– – PVA-stabilized emulsion stability 229–236
– – water-in-oil emulsions stabilized with A-B-A block copolymer 240–245
– flow curves of emulsion 56
– interface, crystalline phases 46
– macroemulsions 2, 29
– nanoemulsions 200
oil slick dispersions 5
oil volume fraction effect, on emulsion rheology 226–229
online droplet size measurement techniques
– direct imaging 115–118
– laser systems 112–115
– sound systems 115
opsonization 202
optical microscopy 189
orthokinetic stability 185–186
oscillating drop and bubble pressure analyzer (ODBA) 99, 100, *101*, 104–107
osmotic free energy of interaction 12
osmotic repulsion, *see* mixing interaction
Ostwald ripening 4, 44–45, 185

p

paraffin emulsions 169
– formation and characterization 178–181
– industrial applications 170
– particle size control 181–185
– preparation 172–174
– properties 170–172
– stability 185
– – as function of electrolytes 189–195
– – as function of Freeze–Thaw cycles 186–189
– – as function of time under shear 185–186
– surfactant systems used in formulation 174
– – phase behavior 175–178
particle vision and measurement (PVM®) 117
phase Doppler anemometry (PDA) 112

phase inversion 4, 47–48
phase inversion composition (PIC) 200
phase inversion emulsification methods 200
phase inversion temperature (PIT) 29–31, 200
photolithography 85
photon correlation spectroscopy (PCS) 240
physical models 161
Pickering emulsions 2
pluronics 1, 210
Pluronic™ unimers 202
polarized optical microscopy (POM) 175
poly(dimethylsiloxane) 82
polydispersity 77, 178–181
poly(ethylene oxide) (PEO) 210–211, *217*, 239, 242
poly(12-hydroxystearic acid) (PHS) 216, *217*, 242
poly(lactic acid) (PLA) 91
polymeric nanoparticles 199–205
polymeric surfactants 1, 209
– adsorbed layers, and droplets 212–213
– – elastic interaction 214–216
– – mixing interaction 213–214
– emulsions stabilized by 216–219
– – W/O emulsions stabilized with PHS-PEO-PHS block copolymer 219–220
– general classifications 210–212
polymer layer overlap *213*, 214
poly(methyl methacrylate) (PMMA) 86
poly(propylene oxide) (PPO) 210–211
poly(vinyl alcohol) (PVA) 210
– emulsion rheology stabilized with 226
– – emulsions stabilized with A-B-A block copolymer 236–240
– – oil volume fraction effect 226–229
– – PVA-stabilized emulsion stability 229–236
– – water-in-oil emulsions stabilized with A-B-A block copolymer 240–245
poly(vinyl pyrrolidone) 210
power draw, in rotor–stator mixers 144–145
premix membrane emulsification 79–80
profile analysis tensiometer (PAT) 99, *100*, 104–105
protein films 51–53
pseudoplastic (shear thinning) system 221
pseudoternary water/mixed surfactant system 177
pulsed drop method 50
pulsed-field gradient (PFG) 119
pulse-echo technique 115

r
reflectance technique 119–120, 123–124
refractive index 121
residual shear viscosity 221
Reynolds number 19, 68, 71
rigidity modulus 226
Ross mixers 132, 146
rotor–stator mixers 127
– advanced analysis of emulsification/dispersion processes in 152–153
– – drop size distribution prediction during emulsification 160–163
– – velocity and energy dissipation rate 153–160
– classification and applications 128–129
– – batch mixers *129*, 132–133, *134*
– – colloid mills *129*–130
– – design and arrangement 133–136
– – in-line mixers *129*, 130–131, 145, 146, 147, 148, 157–160
– – operation 136–137
– – toothed devices *129*, 131–132, *134*
– engineering description of emulsification and dispersion processes 138
– – average drop size in liquid–liquid systems 145–147
– – drop size distributions and average drop sizes 138–140
– – drop size in liquid–liquid two-phase systems 140
– – maximum stable drop size in laminar flow 141–142
– – maximum stable drop size in turbulent flow 142–143
– flow characterization 143–145
– possible arrangements, for batch processing *135*
– products manufactured using *128*
– scale-up rules 147–152

s
Sauter mean diameter 140, 143, 146, 150, 151
Schultz–Hardy rule 194
shear flow, simple 141
shear stress 66, 79, 81, 143–144
shear viscosity measurement, interfacial 49–50
Shirasu porous glass (SPG) 80
Silverson mixers 131, 132, 133, 134, *136*, 145, 146, *148*, 157
single-pass processing, of rotor–stator mixers 136, 137

SINTERFACE Technologies (Berlin, Germany) 99
small-angle X-ray scattering (SAXS) measurements 175, 178, 183, 184
Smoluchowski equation 189, 192
sodium dodecyl sulfate (SDS) 1, 16, 24
solvent evaporation method 201, 204
sound systems 115
spatial filtering velocimetry (SFV) 113
stability curves 141–142
steady state measurements 220–222
stereomicroscope 116
steric interaction free energy 213
steric repulsion 12–13
steric stabilization 40, 209–210
– energy–distance curves 215
– oil-in-water (O/W) emulsion rheology stabilized with poly(vinyl alcohol) 226
– – emulsions stabilized with A-B-A block copolymer 236–240
– – oil volume fraction effect 226–229
– – PVA-stabilized emulsion stability 229–236
– – water-in-oil emulsions stabilized with A-B-A block copolymer 240–245
– polymeric surfactants
– – adsorbed layers and droplets 212–216
– – emulsions stabilized by 216–220
– – general classifications 210–212
– rheological technique principles
– – constant stress (creep) measurements 222–223
– – dynamic (oscillatory) measurements 223–226
– – steady state measurements 220–222
Stokes–Einstein equation 41
Stokes law 36, 204
straight-through microchannel arrays 88–89
strain sweep 224
stroboscope 116, 117
surface charge density 191
surface excess 14
surface potential calculation methods 191–192
surfactants 17, 18, 80–81
– systems, used in formulation 174
– – phase behavior 175–178

t

thermodynamics, of emulsion formation and breakdown 6–8
thickeners 38, 220
T-junction 82
toothed devices *129*, 131–132, *134*
transmembrane pressure and wall shear stress 81
transmission electron microscopy (TEM) 204
turbulent flow, maximum stable drop size in 142–143
turbulent inertial (TI) 71
turbulent viscous (TV) 71

u

ultrasonic spectroscopy 115
Ultra-Turrax mixers 131, 226

v

van der Waals attraction 4, 8–9
van der Waals potential 190, 193
velocity and energy dissipation rate 153–160
viscosity ratio 146
– modified 69
viscosity–volume fraction
– curves 57
– relationship 68, 241
volume restriction interaction, *see* elastic interaction

w

water/ionic surfactant system 175
water/nonionic surfactant system 175
water-in-oil (W/O) 203
– emulsions, stabilized with A-B-A block copolymer 240–245
– macroemulsions 2, 29
– – flow curves of emulsions 56
Weber number 20, 140
Weissenberg number 69
Weissenberg rheometer 79
wide angle X-ray scattering (WAXS) 170
Winsor concept 31

y

yield stress 221, 226, 229
Y-junction 83

z

zero shear viscosity, *see* residual shear viscosity
zeta potential 189–190, 191–192
Zwitterionic surfactants 81